BLUE HOUSES

Sustainable Homes

BLUE HOUSES
Sustainable Homes

Cristina Paredes Benítez

LOFT

Blue Houses. Sustainable Homes

Editorial coordinator:
Simone K. Schleifer

Assistant to editorial coordination:
Aitana Lleonart Triquell

Editor:
Cristina Paredes Benítez

Texts:
Loft Publications

Art director:
Mireia Casanovas Soley

Design and layout coordination:
Claudia Martínez Alonso

Cover:
María Eugenia Castell Carballo

Layout:
María Eugenia Castell Carballo, Guillermo Pfaff Puigmartí

© 2011 **Loft Publications**
Loft Publications, S.L.
Via Laietana, 32, 4º, of. 92
08003 Barcelona, Spain
T +34 932 688 088
F +34 932 687 073
loft@loftpublications.com
www.loftpublications.com

ISBN 978-84-92463-93-0

Printed in Spain

Active and passive supply and energy saving systems.

 Use of recycled, recyclable or prefabricated materials.

 Systems to save or reuse water.

 Architectural elements that minimize the environmental impact on the setting.

INTRODUCTION

In this book we aim to make our readers better acquainted with sustainable architecture. All kinds of options are presented, since green architecture is not only a question of having a house built without CO_2 emissions or one that is completely self-sufficient with respect to electricity or water consumption. Being an environmentalist starts by adopting an attitude and implementing changes in one's behavior and daily routine. While leaving the car at home and using public transport or cycling to get about are practices that contribute to the sustainability of the planet, there are other minor activities performed in our homes that can also be helpful. That is to say, sustainability and ecology also involve using energy efficient compact fluorescent light bulbs (CFLs), installing water flow regulators on all the faucets in the house, and increasing insulation so as to avoid wasting heat in winter.

Architecture is considered to be green when it takes account of a number of different factors during the process of constructing a house and the impact a building will have throughout its life cycle: from the moment it is designed, erected and used until it is pulled down at the end of the process. Residential green architecture also takes account of the specific needs of the people that will live in each house.

Some trends are based on new prefabricated materials. Mass production of such materials leads to lower costs, along with faster and easier assembly. This means that savings are also made in greenhouse gas emissions generated by the excessive transportation of materials and, in the case of housing in the country or in mountain regions, the environmental impact on the natural surroundings is softened. Other solutions are based on a type of architecture that has traditionally been more respectful of the environment. In these cases, the sites are examined to ensure that the homes are oriented in accordance with their needs and adapted to the local climate, so that they can take advantage of the hours of sunshine and the air currents, resulting in enhanced energy efficiency. Furthermore, science and technology have developed active systems to enable rainwater to be reused or wind and solar power to be generated. All these topics will be discussed in greater detail in the following text, which explains the most common systems—both active and passive—for achieving an environmentally friendly home.

The following projects demonstrate a large range of sustainable and ecofriendly resources: from recovering traditional architectural solutions such as cross ventilation or the use of natural light to new systems of insulation and self-sufficient energy supplies. The projects, located in various parts of the world, are examples of environmental trends in architecture, a movement that is gaining more and more followers with each day that passes. These ideas and proposals will give our readers a general idea of a type of architecture that is real and in a constant development. *Blue Houses* relates residential architecture with significant and current topics about climate, water, energy and cities, all from an ecological and sustainable point of view.

THE ABC OF SUSTAINABLE ARCHITECTURE

A BRIEF INTRODUCTION TO THE MOST COMMON TERMS IN GREEN ARCHITECTURE

For several decades now the concepts of ecology and sustainability have generated highly influential currents of opinion in our society, invading all areas of our life. Nowadays no one goes against the theories of global warming or the ever more pressing need to reduce greenhouse gas emissions. In the field of architecture, which is the subject that interests us here, significant changes and developments are being done. However, reality tells us that we still have a way to go. The percentage of sustainable buildings is still very low and in open conflict with the increase in the number of building projects practically all over the world. In the field of residential architecture alone, there is an increase in the number of homes as a result of the population growth. This in turn leads to greater consumption of raw materials and rising environmental costs for their transportation.

Much more participation is still required on the part of everyone involved, from architects and building contractors to manufacturers of materials and end customers, if what we hope to achieve is a quality building that is also environmentally friendly and sustainable at the same time.

This book aims to bring the current trends in sustainable architecture to a general non-specialist audience. Most of us might feel completely overwhelmed by the vast quantity of new concepts such as ecological footprint, environmental impact, recyclable and recycled material, thermal mass, biomass, etc. In the course of the following pages all these terms will be clarified, together with many more that can help the reader gain a more precise knowledge of this subject. They will be arranged in three main sections: building systems, materials and water savings.

BUILDING SYSTEMS. THE ROAD TO RESPONSIBLE ARCHITECTURE

In green architecture building systems are divided into two large groups known as passive and active systems. The first group covers all options that place emphasis on improving the energy efficiency of buildings devoid of elements that generate artificial power. The second group consists of systems that obtain power by harnessing so-called renewable energies.

Passive systems. Reducing energy consumption is in our hands

Buildings constructed with passive systems are also known as bioclimatic buildings, i.e. they use non-artificial elements to heat or cool the home.

The first thing that needs to be done is to look at the orientation of the property. Normally buildings are oriented towards the south, or towards the east, to maximize the use of daylight hours. This lowers electricity consumption and the heat generated is used to reduce conventional heating requirements. The concept of thermal mass harnesses these solar radiations: some building materials such as concrete store the heat and release it hours later. Thus, if we let the sun enter a living room with concrete walls or flooring, these will radiate heat at night, making it unnecessary for it to be generated.

Depending on the climate in the area, it may be necessary to vary this orientation or to add elements that will offer optimum HVAC conditions for the house throughout the whole year. A residence built in a country with a warm or even tropical climate might require a different orientation. The slope of the roof and the eaves can be calculated by considering the differences in the angle of the sun's rays in winter and summer, thereby avoiding excessive exposure during the warmest months of the year.

Another simple solution enabling savings to be made in energy is the Trombe wall. This system entails painting a wall a dark color and installing a pane of glass in front of it so as to create an air chamber. It is more effective when installed in south-facing walls in the northern hemisphere and in north-facing ones in the southern hemisphere. Openings in the wall will enable warm air, which is less dense, to rise and penetrate the building. An opening lower down will allow the cold air to exit. Moreover, if these openings can also be closed at will, the house can also be ventilated or cooled thanks to the exit of warm air in the summer.

The green roof is another of the systems used to help regulate the temperature inside the house. There are several advantages that make this covering a good insulation system when compared with traditional roofs. The most obvious benefits are the management of rainwater and reduced energy costs. In addition, green roofs or covers also reduce the heat island effect found in large cities. Thanks to their breathability and the shade provided by plants, the use of such roofs cools the building and helps to reduce the temperature in both housing and cities. Thus, when the temperature inside the house falls, this reduces the need for air condi-

tioning. Furthermore, this system extends the life of the roof and helps ensure good sound-proofing, which is necessary in noisy environments such as cities or locations near airports or industrial areas. If a green roof is installed, it should be borne in mind that a certain amount of maintenance will be required, and hence there will need to be safe access to the roof. Apart from green roofs, the use of plants such as creepers on the walls of the property also contribute to energy savings for the same reasons.

The ventilated façade is a system for insulating vertical walls consisting of several layers with ventilation in between. These exterior enclosures are composed of a number of sheets, with the outside finish or siding being a composite of various materials such as natural stone, metal or plastic paneling, etc. These top layers are fitted in place leaving a slight gap to ventilate the chamber through the joins that have been created. In this way it is possible to create a powerful form of thermal insulation attached to the inside of the wall—and furthermore with constant ventilation, thereby preventing any condensation. Should any water get in through the joins, it could be evacuated via the inner surface of the cladding or siding. The main advantage of this type of façade is that thermal bridges disappear, these being zones where heat is transmitted more easily, since they contain heat conductive materials, such as metal.

Installing double-glazing in the windows brings substantial improvements in both thermal and acoustic insulation. If the window frame is able to cause a break in the above mentioned thermal bridge, the insulation will be that much more effective.

If the house has a shaded inner courtyard, it might be enough to incorporate water and vegetation to help cooling the atmosphere. The presence of water, whether in the form of a fountain or a small pond, will cool the air since, as it evaporates, the energy is absorbed. In the case the residence has land around the building, some trees and vegetation will provide shade and shelter from the wind, apart from offering more privacy.

Another of the strategies recovered for sustainable architecture is cross ventilation. If the orientation of the house has taken account of the air currents and the design of the disposition of the windows, doors and verandas, these currents should be used to cool the house, and thus the air conditioning system may prove to be superfluous, making it possible to reduce the electricity consumption. Even if the design is not entirely appropriate, the openings in the outer walls may generate currents of air that will help to cool the interior of the house.

Finally, the installation of awnings, shutters and eaves will help regulate the temperature inside the house. The shutters provide protection from the heat and create a second layer of insulation, which is useful in both summer and winter. Eaves provide fixed shade and awnings can be lowered or taken up as the need dictates. Blinds have the same function, as well as offering the occupants more privacy.

Active systems. How to harness the power of nature

Active systems are those that harness and use the power generated by inexhaustible sources, i.e. renewable energies that are obtained from natural resources. The best known are solar, wind and thermal power, but there are other types such as biomass, geothermal, hydro or tidal power. These are also known as alternative energies.

They are not all applied to residential architecture but sufficient technologies have been developed to ensure that a house can be equipped with every amenity without it being necessary to resort to conventional forms of power (which are generated by fossil fuels and nuclear energy). Thanks to these solutions, it is possible for a house to be sustainable and self-sufficient.

One of the energies that is most widely known and used is solar power. Solar radiation provides far more energy than is consumed on the planet, and the systems that have been developed employ it to avoid using conventional forms of energy. There are two types of panels —solar collectors or thermal solar panels, which use solar power and transform it into heat, and photovoltaic solar panels, which transform it into electricity. Both methods use different technologies, but are supplied by the same source.

Thanks to various physical principles, thermal solar panels enable current consumptions of gas, oil, or any type of conventional energy used for heating and hot water to be reduced or eliminated altogether. They have the advantage that in most cases the installations only need to be modified rather than changed completely. The system does not generate CO_2 emissions and the useful life of the boiler is extended, since the number of stops and starts is reduced and combustion improved. The system aims to reduce the operating hours of the boiler, and hence fuel consumption.

Photovoltaic solar panels convert solar radiation into an electric current, which can then be used to power any appliance. Due to global warming and the other effects of pollution, there is greater environmental awareness on the part of society and this new mindset has translated into a greater demand for this type of system.

To manufacture these panels or cells, advanced complex technology is used that is different from that used for thermal solar panels, which is available to many more businesses. The operation of this type of panel is based on the photovoltaic effect. We shall not go into details here; we shall, however, attempt to summarize the idea to make it easier to understand the basics of this technology. The photovoltaic effect occurs when solar radiation is incident on suitably treated semiconductor materials, which generate fields known as a p-n junction. When exposed to the sun's rays, these produce an electric current. The smallest module of this semiconductor material with a p-n junction and thus with the capacity to produce electricity is called a photovoltaic cell. These cells are combined to attain the power and voltage required. A photovoltaic panel is formed by these cells, which are suitably protected and mounted in the appropriate frames.

There are various types of PV panels, depending on materials used (among others, monocrystalline or polycrystalline pure silicon, cadmium telluride or amorphous silicon), form (reflective, tile format, bifacial panels, etc) or the solar tracking systems available (one or two axis, or static support). The needs of each residence and the type of climate will determine the type chosen. Attention should also be paid to the amount of power required in the home when calculating the number of panels that will need to be installed. If the space available on the property is not sufficient, the electricity consumption obtained from unsustainable methods might at least be reduced.

In short, a domestic system for generating electricity using photovoltaic panels can be based on three possible solutions—an autonomous circuit for small installations, a support system for electricity consumption at the property, and a production system connected to the mains.

There is a third type of panel—the so-called composite panel—which incorporates a circuit of water-heating tubes. On the plus side, it has all the advantages of the other two types of system. However, its expensive cost is obviously a real drawback.

Properties that only use clean electric power are known as *off the grid homes*. These homes are completely disconnected from the conventional electricity grid and receive their power supply by harnessing active or passive strategies.

Another of the terms often mentioned when speaking about active systems is biomass. Biomass is the organic material generated through a forced or spontaneous biological process, which is used as a source of energy.

Quite a common error is the use of the term biomass as a synonym for the useful energy it can generate. Useful energy can come from the direct combustion of biomass (wood, nutshell, etc) and from fuels derived from it through physical or chemical change (like the methane gas generated by organic waste, for example), although there is always some loss in the case of the latter. Although biomass can be classified in several ways, here we have chosen the most widely accepted categories. Thus, biomass can be divided into four different types—natural, wet and dry waste, and energy crops.

Natural biomass is produced by nature without any human intervention. The main disadvantage is the expense of obtaining it and its transportation from source to households that will use it as fuel; it may well be that harvesting this resource proves to be unfeasible from an economical point of view. On the other hand, clearing forests of this type of waste would be of benefit to the environment, by preventing the build-up of forest mass that might increase the severity of an accidental fire. Residual biomass (both wet and dry waste) is composed of

the waste generated by farming, livestock and forest activities (timber, and food and agricultural industries, among others). Such by-products, which are still usable, are for instance wood and grass waste, sawdust, almond shell, the remains left after pruning fruit trees, etc. Last of all, we have the biomass that comes from the so-called energy crops, which are crops generated for the sole purpose of producing biomass that can be turned into fuel.

The energy harvested from biomass that can best be used in residential architecture is the direct combustion of natural biomass in stoves that burn wood or pellets, or that use fuel for heating that comes from energy crops, although this can also generate CO_2 emissions.

Another active system used in residential architecture is geothermal power, which is obtained by harnessing the heat present in the subsoil. With this type of energy, it is possible to power a home HVAC system and heat the water supply in an environmentally friendly fashion. The geothermal HVAC system cedes or extracts heat from the earth through a set of collectors buried in the ground with a water and glycol solution circulating through them.

The geothermal HVAC system operates as follows: to cool a building in summer, the system transmits the excess heat from the interior of the building to the subsoil. In winter, the geothermal installation heats the building following the reverse procedure, extracting heat from the ground to transmit it to the building by means of collectors. This technology can be installed in any type of building, apartment block, second home, etc—even in buildings that are already standing.

There are several advantages to this system: this type of energy can prevent the homeowner from becoming dependent on energy from an external source. In addition, the waste generated is minimal and causes a smaller impact on the environment. It improves the aesthetic appearance of the building since there are no external elements on outside walls and roofs, which also amounts to saving space. It is silent and compatible with other renewable energies. While few and far between, the drawbacks are that it is not available everywhere, and cannot be transported. It should be remembered that small emissions of hydrogen sulfide may be generated, which can be detected by its smell of rotten eggs.

Another active system available is the one consisting of micro wind turbines. There is no single or conventional classification for this type of installation, but as a rule they are similar to the larger versions of wind turbine found on wind farms. Nevertheless, the technology of these microturbines is different and their usual purpose, unlike that of the larger machines, is to produce energy for personal consumption. Their application is really useful, particularly in places where there is no electricity supply, such as isolated spots, rural areas with schools, health centres, isolated tourist infrastructures, etc.

However, they are also very useful for providing green electricity, even in places where an electricity supply does exist. Direct application at the point of consumption avoids energy loss deriving from conversion and transportation. Furthermore, microturbines can produce approximately 30% of the energy consumed by each household and do so with less visual impact, lower costs, greater efficiency and greater sustainability. The majority of micro wind turbines have a horizontal axis and three blades, although it is also possible to find them with one or two blades. There are also micro wind turbines with a vertical axis.

The main disadvantage of these systems is the acoustic impact generated by the turbines. This impact is mostly caused by horizontal-axis wind turbines. This is largely due to the fact that the spin axis is parallel to the ground, which produces a lot of friction because of the torque. The solution, if the turbine needs to be installed near the home, is the use of vertical-axis microturbines. It is thought that the operating noise produced by these microturbines is 0.5% less than that of the horizontal version.

The implantation of these machines will be subject to standard regulations and technical specifications, setting out the maximum distance for the location of these installations with respect to the point of consumption, the possibilities of using and combining them with other renewable (or conventional) energies, the power required and permitted for each household, etc.

ECOMATERIALS. HOW TO USE AND REUSE THE MATERIALS THAT NATURE GIVES US

One of the main premises of green and sustainable architecture is the return to architecture's own origins. One of the aspects most associated with this philosophy is the choice and use of building materials.

The use of locally sourced raw materials is always less costly, not only because of the economic value of the material itself (if it is sourced locally, it is likely to be more plentiful in that particular geographical area and therefore cheaper), but also because of the environmental costs incurred in transporting it. Choosing stone from European or Asian quarries to clad a residence in the United States may be a very aesthetically pleasing option, but it is not very sustainable, since on top of fuel consumption and the CO_2 emissions deriving from its transportation, we also have to add the manufacturing costs along with that of the material itself. Tropical timber can be very resilient, but might not be the best option for building a home in the Nordic countries. The use of time-honored techniques that have been around for centuries, such as adobe, may well be a sustainable option for building a house, if the stability and safety of the home can be guaranteed.

Another parameter that should be considered is the use of so-called natural materials. Grouped together under this umbrella term are materials that require few manufacturing processes for their production. Each of the artificial processes a material goes through from its extraction or initial production until it leaves the factory has a cost in terms of energy and hence environment, and therefore the use of materials requiring few processes will always be more sustainable.

There are a great many natural materials. Fortunately, the industry recognizes the advantages inherent in these types of materials, and the greater the offer, the more likely it is that the end users will demand and use materials that are more respectful of the environment. Among the best known are wood and bamboo. Treatments can be very diverse, ranging from the absence of varnish to the use of lacquers that are respectful of the environment. Straw, which is used as insulation, is also a readily available natural material that can be utilized in construction. Other insulating compounds, such as *cannabric*, do in fact follow manufacturing processes that guarantee appropriate, approved specifications for use in construction, but are obtained from industrial hemp fibers.

In the category of green building materials we also find certified materials. One of the main ones is certified wood, that is to say, wood harvested from controlled forests, using methods that are respectful of the environment. Two of the most important associations that guarantee this controlled source are the FSC (Forest Stewardship Council) and the PEFC (Pan European Forest Council), the latter being managed by the producers themselves. This

ensures that no unprotected forests are used, which guarantees the sustainable management of this raw material and helps slow the loss of green areas on the planet.

There are some techniques that use natural materials, such as rammed earth, which mixes local soil with a small amount of concrete. This mixture is poured into the formwork and pressed to obtain a building that is just as safe and stable but using soil that has been sourced from the same building site.

To finish, the use of marble or stone will be sustainable if their properties are adapted to the type of home in question and if they are sourced locally.

Apart from those already mentioned, there are other types of components that can be used in sustainable architecture: recyclable and recycled materials. The first of these concerns materials that can be reused with ease, either directly or after being subjected to a number of processes, and the second group are composites produced from waste or other components that are given a second use. It is important that these materials do not generate waste either during production or when they finish their useful life.

Recycled materials are partly made with existing materials that have been modified by an industrial process. Such materials can be recycled again and again, as many times as necessary, once the home's useful life has come to an end.

Wood and many of the natural materials discussed above are materials that can be given a second life. In other words, they are recovered materials—timber that can form part of the beams in a new house, stone that can be used for a new piece of work nearby, etc. The reuse of resources satisfies several principles of sustainable architecture—local materials are used, savings are made in transportation and therefore in CO_2 emissions, and resources and materials are not used in an uncontrollable fashion. From an aesthetic viewpoint, the recovery of stone, timber or cladding from other buildings might lend character to the new house. It is also possible to recover other waste materials, from industrial or urban sources, to form part of some of the elements in a house such as railings and banisters, window frames, etc.

WATER SAVINGS. TOWARDS A NEW CULTURE OF RESPONSIBLE CONSUMER HABITS

Water is an essential element for human development. It is also true that the commodity is in short supply and cannot afford to be squandered. Global warming brings drought and desertification in its wake, and thus it is increasingly urgent to find a rational way to deal with water management. For domestic purposes, it is normal for the water supply system to always use drinking water, regardless of use. Such expenditure is unsustainable and totally uncalled for—and not only from the standpoint of green architecture. Squandering this resource damages this environment. We therefore need to adopt the so-called new water culture.

There are several ways to save water in a home. One of these might be to inspect the pipes to make sure they are in good condition and there are no leaks in the system. Others are simple and practical solutions that allow savings in levels of consumption and also on the bill, ranging from taking a shower instead of a bath to installing water flow regulators in all faucets. It is also common to install cisterns with stoppable flushing mechanisms and to purchase household appliances with greater energy efficiency i.e. class A.

Drip irrigation systems and timers can be installed in gardens, which will also avoid excessive waste. The choice of native vegetation allows a balance between the needs of the plants and the amount of rainwater available. Other systems that are very popular nowadays for improving domestic water management involve the treatment of gray water and rainwater.

Gray or wastewater refers to the water generated by household chores and routines: the water used in washing machines and dishwashers, and the water from showers and bath tubs. They differ from sewage or black water because they do not contain the bacteria *Escherichia coli*. Gray water usually decomposes faster than black water, and contains less nitrogen and phosphorus. However, it also has a percentage of black water that includes various pathogens. Nevertheless, such water can be simply treated for reuse in toilets, which means significant savings in water and avoids using drinking water in situations in which it is not absolutely necessary. However, we should consider the fact that untreated gray water cannot be used to flush the toilet since it might generate bad odours and stains if left for more than a day. There are systems that can treat this water, by decanting it, for example, or using biological filters. This type of water is useful for watering gardens. Although quite low, its content in phosphorus, potassium, and nitrogen is beneficial to plants and also prevents such elements from polluting the rivers and lakes.

The systems available for harnessing rainfall harvest the rainwater, as the name suggests, for use for other purposes, such as watering gardens and cleaning houses, both in and outside. Rainwater is also used for dishwashers and washing machines. Some of the points in its favor are that, since it is rainwater, it is much softer, and therefore much less detergent is needed. A good circuit for collecting rainwater can be achieved by installing a pipeline that runs from the roof down to the water tanks. The tank capacity should be decided according to the rainfall for the geographical area in which the home is located and the needs of the occupants. This will ensure that water can be stored for the uses required: watering the garden or vegetable patch, for household appliances, etc. A good water collection system should be simple and require very little maintenance. Several factors should also be avoided that might affect the quality of our water, such as dirt, light or too much heat, which means that the tanks need to have adequate protection.

There are some systems that use the rainwater stored in these tanks while stocks last, but, where necessary, can connect to the normal mains, thereby ensuring a constant supply. It is essential to have a good filter at the entrance to the tank if the water is to be put to domestic use. This would not be so important if the water were only meant for gardens and other outdoor purposes. The tank material must also be environmentally friendly. Therefore, tanks made of PVC or plastic reinforced with fiberglass are not recommended. The ideal temperature should not exceed 12 ºC, and thus it is recommended that tanks be buried underground. The pump is one of the most important elements and should therefore be tough and long-lasting. The ones most recommended are those made of polyethylene, which are cheap and resilient. The water intake valves in the pipes running from the rainwater tank should be labeled and easily distinguishable from all other stopcocks. Among the disinfecting systems that guarantee the microbiological potability of the water, one of the best is the one that uses ultraviolet rays.

A final system that could also be applied to sustainable domestic architecture might be the use of water emptied out of swimming pools. This follows the same principle as the system for recovering rainwater and could also be used for external cleaning purposes and watering garden areas.

Cristina Paredes Benítez

Sustainability throughout the entire home

La Bonne Maison | Coste Architectures | Prouais-sur-Opton, France
© Vincent Fouin

This latest model for a prefabricated home was the brainchild of the famous photographer Yann Arthus-Bertrand. The design is by Agence Coste Architectures, but the actual construction is the responsibility of the Geoxia consortium (Maisons Phenix). These houses began to be marketed in 2008. From the outside, the residence appears to be made up of fairly classic cubic forms: a rectangular shape and gable roof. The outer cladding on the walls is wood.

From the perspective of sustainable architecture, the house offers a multitude of solutions for greener living. The first is its optimum orientation to the sun, since 72% of the natural light penetrates the building on its southern side. Thanks to this disposition and to all the other bioclimatic features, the property does not need to be heated by electricity. To achieve optimum insulation and prevent heat loss, the project integrated the total elimination of thermal bridges and triple-glazed windows.

In addition, the walls were erected with a double R-value in terms of thermal resistance, in accordance with French technical regulations.

The property incorporates thermal solar panels, located on the roof, which provide over 50% of the energy required to heat the water used in the home.

An earth tube (earth-air heat exchanger) buried in a trench at a depth of 2.5 m feeds a ventilation system with air at a constant temperature of 14 ºC (with a variation of ±2 ºC). The air passes through a double-flow thermodynamic system and is distributed by two diffusers that help to renew the atmosphere inside the house throughout most of the year. A wood-burning stove has also been installed to enhance the heating system.

The energy consumption for the entire house is 12 kW/m^2 per year, whereas in normal conditions, without taking any measures to increase energy efficiency, it would amount to roughly 70 kW/m^2 per year. The savings in energy, which are calculated to be around 80%, and in water, estimated at about 38%, translate to a reduction in CO_2 emissions into the atmosphere of some 1250 kg a year—figures that are by no means negligible, showing that reducing emissions is a viable proposition.

 Solar panels for heating water and natural ventilation system.

 Triple-glazed windows, walls with thermal resistance with a double R-value.

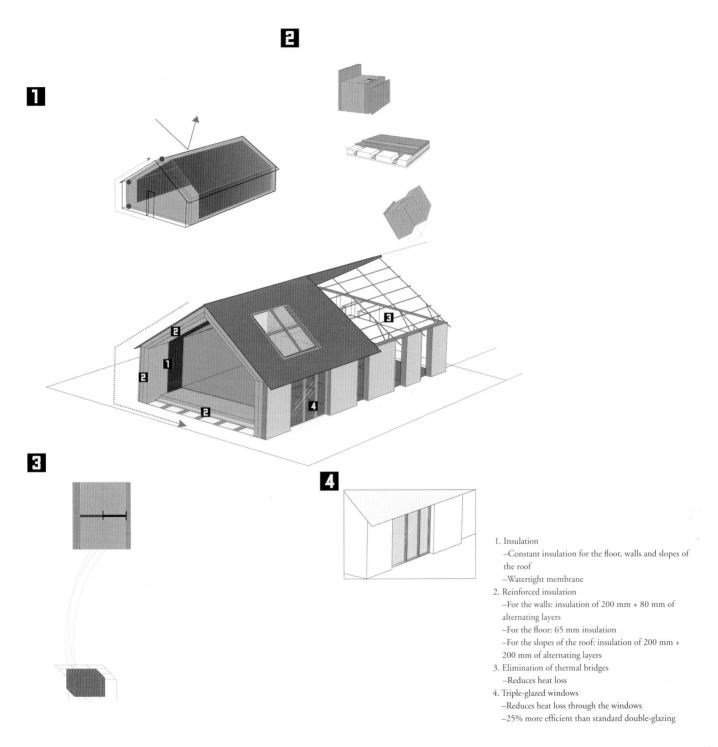

1. Insulation
   –Constant insulation for the floor, walls and slopes of the roof
   –Watertight membrane
2. Reinforced insulation
   –For the walls: insulation of 200 mm + 80 mm of alternating layers
   –For the floor: 65 mm insulation
   –For the slopes of the roof: insulation of 200 mm + 200 mm of alternating layers
3. Elimination of thermal bridges
   –Reduces heat loss
4. Triple-glazed windows
   –Reduces heat loss through the windows
   –25% more efficient than standard double-glazing

The wood cladding and thermal solar panels are the most striking features on the outside of the house, which follows a traditional design, with rectangular shapes and a gable roof. The disposition of the house maximizes heating and lighting.

Axonometric view

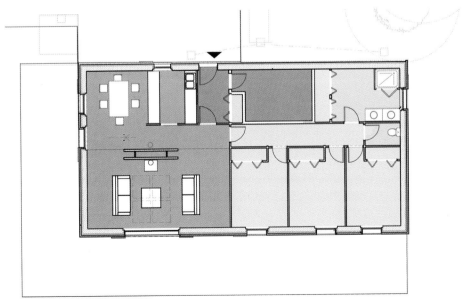

Ground plan

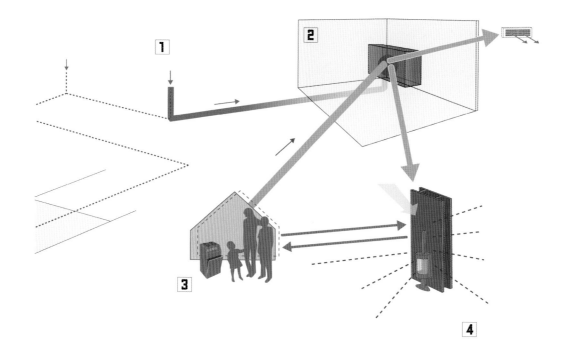

1. Passive geothermal power: an earth tube recovers air from the outside to enable it to enter the house at a constant temperature (14 °C ±2 °C).

2. This is an innovative, non-conventional heating system. It acts like a double-flow thermodynamic ventilation system that ensures renewal of the air inside the home. The incoming air is preheated by the outgoing air.

3. Energies normally left untapped. All electrical appliances and the human body release heat. The insulation of the property is such that it is possible to take this form of heat into account in the final analysis.

4. Wood-burning stove and wall with thermal inertia properties. In order to create indoor comfort, a wood stove has been installed that uses energy drawn from the biomass, which in this case is wood.

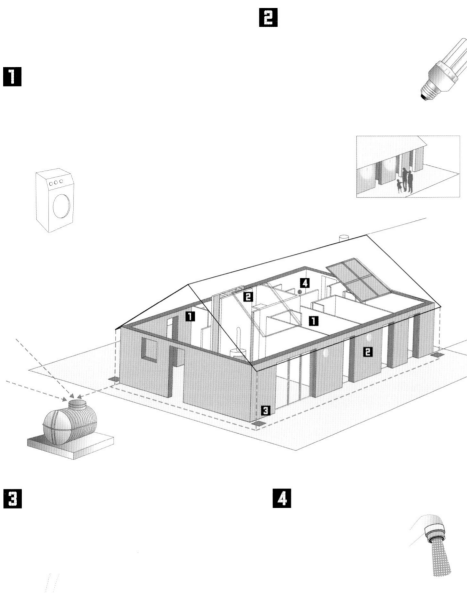

1

2

3

4

1. Electrical household appliances
   –31% saving
   –The choice of household appliances should be made on the basis of their energy efficiency. Category A++ appliances have been selected.
2. Lighting
   –75% saving in energy when compared to conventional lighting
   –Use of energy-saving compact fluorescent light bulbs (CFLs)
   –Motion detectors in external lighting
3. Recovery of rainwater
   –Rainwater is harvested in a tank buried underground and reused as gray water: for watering the garden and washing the car. This represents a 25% saving in water consumption.
4. Water savings
   –This system replaces part of the water with air, thereby obtaining savings with the same apparent flow of water. This device contributes to energy savings because it also reduces hot water consumption.

The thermal solar panels are used to heat the water supply and the house itself. In this case, over half the water used in the home is heated using this method. This leads to significant energy savings and avoids CO_2 emissions deriving from burning fossil fuels.

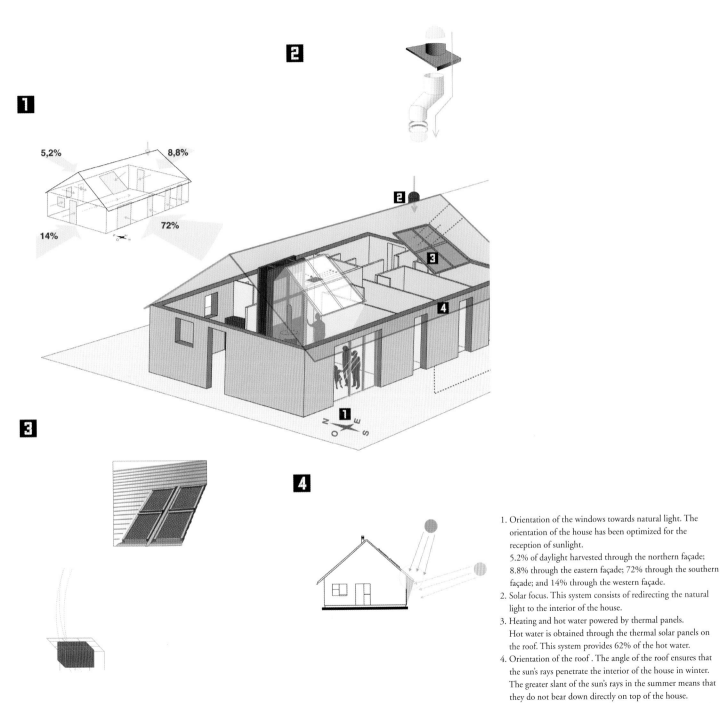

1. Orientation of the windows towards natural light. The orientation of the house has been optimized for the reception of sunlight.

   5.2% of daylight harvested through the northern façade; 8.8% through the eastern façade; 72% through the southern façade; and 14% through the western façade.

2. Solar focus. This system consists of redirecting the natural light to the interior of the house.

3. Heating and hot water powered by thermal panels. Hot water is obtained through the thermal solar panels on the roof. This system provides 62% of the hot water.

4. Orientation of the roof . The angle of the roof ensures that the sun's rays penetrate the interior of the house in winter. The greater slant of the sun's rays in the summer means that they do not bear down directly on top of the house.

Prefabricated architecture as a means of obtaining a sustainable home.

Wood and steel house | Ecosistema Urbano Arquitectos | Ranón, Spain
© Emilio P. Doiztua

Located in the rural setting of Asturias, Spain, this property represents a complete modernday overhaul of the models of architecture that are traditional in the area: raised granary and glazed gallery, use of wood in the structure and enclosure. The in- and outside are both covered with tongue and groove wooden boards.

To respect the site on which the property is located, the building was only anchored to the ground at four points, thereby achieving a compact result. The fact that this construction is prefabricated reduces the building times and waste materials that would be generated by a conventional building.

The structure is formed by a prism that is distorted and displaces the southwest corner. These forms are not just a whim of the architects or the owners of the property: this design does in fact allow more sunlight to penetrate the interior. Furthermore, the upper plane of the roof leans against the hillside making it easier to evacuate rainwater. On the southern side, bathed by a large amount of sunlight, the façade is fully glazed. The bioclimatic features are guaranteed by virtually windowless façades that prevent any loss of heat. Vents are included for the entrance and exit of air on all sides of the property, so as to create cross ventilation inside the building. Thanks to such measures, which maximize the use of natural resources, the house has no heating or cooling system.

Rather than installing a conventional system of adjustable slats, Ecosistema Urbano Arquitectos manages to create shade by using a system of vents in various positions and combinations that act as a hygrothermal regulator that is better suited and more appropriate to the microclimate existing in the region.

In the opinion of the architects, the house is flexible, adaptable and shareable. Its position and original geometry mean that the building is perfectly adaptable to the climate and orientation of the sun in the area. The environmental features of the project are completed by the prefabricated structure.

 Passive air-conditioning systems for the property.

 Prefabricated materials that can be recycled and dismantled.

The northern façade is protected by an anterior space and a wind-blocking lattice window. On the main façade it is possible to observe the double height of the interior, which has not been designed for reasons of space or composition, but is instead an essential bioclimatic tool for regulating the temperature of the house.

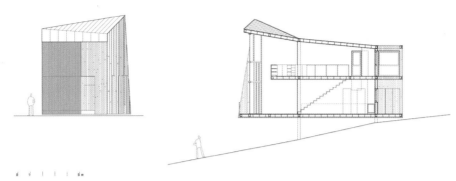

Cross elevation and section

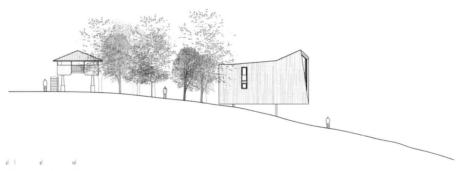

Longitudinal elevation

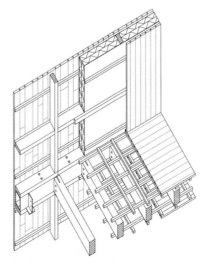

Detail of cladding on the wall and floor

The structure of the building is a composite made of wood and steel. It can also be dismantled and recycled. Its walling uses a combination of two types of wood—North pine and Douglas pine —with two different widths. The outer cladding consists of tongue and groove wooden boards with a thickness of 35 mm.

The ground floor incorporates multiple uses in a single area that can be configured by users in different ways. The first floor can include up to three bedrooms.

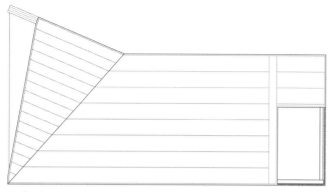

Roof level

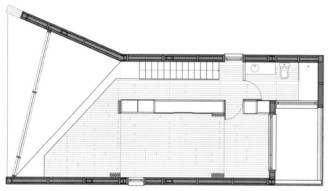

First floor

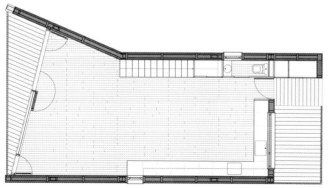

Ground floor

To respect the land where the house is located, the building is anchored to the ground at only four points, so as to achieve a compact result. The fact that this construction is prefabricated reduces the construction time and waste materials that a conventional building would generate.

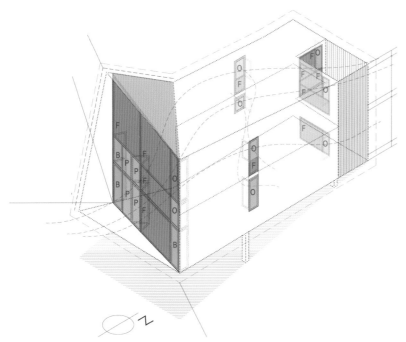

The top diagram illustrates the ventilation generated by the openings. The bottom diagram summarizes various characteristics of this construction: one of the natural materials (wood), light and heat transmission through glass façade, recuperation of traditional architectural techniques, etc.

Diagram of the openings and cross ventilation

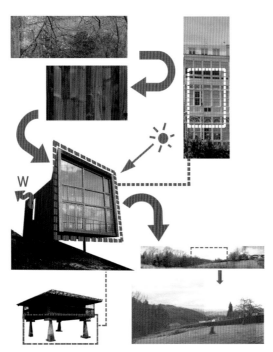

Ideogram of the house design

Adaptation to the natural surroundings

CO_2 Saver | Architekt Kuczia | Lake Laka, Poland | © Tomasz Pikula

Like a chameleon, this little house adapts to the environment and blends with the surrounding countryside. Its outline is symmetric, just like the shape of most animals, although its internal spaces are unevenly distributed as befits the needs of the owners. The design and construction of the property were based on environmental criteria and to this end passive systems were designed and active systems installed.

The shape of the building was intended to help absorb the maximum amount of solar energy. Nearly 80% of the structure is oriented towards the sun, which floods in through the glazed patio on the ground floor. Solar panels have been installed on the roof to provide heating and hot water for the occupants, and more are scheduled to be installed in the future to make the building self-sufficient where electricity is concerned. Another main aim was to obtain good thermal insulation. The outside of the building, comprising three stories, is clad with charcoal-colored fiber cement siding—a material that reduces heat loss. The wooden façade is composed of boards stained in similar colors to those found in the surrounding countryside. The passive and active insulation systems have been reinforced with a ventilation plant equipped with a thermal recovery system.

The project design was also determined by the need to control building costs. The house, which did not cost any more than a conventional home in the same area, implemented ideas and resources that proved to be most appropriate since they enabled savings to be made in terms of construction costs. This was achieved by applying traditional building techniques and using local materials and recycled building materials. Furthermore, the materials used have not been treated chemically and are easy to repair or reuse, which reduces maintenance costs.

A green roof in some areas completes the set of environmental measures that have been integrated in the house to achieve maximum sustainability and optimum adaptation to the surroundings.

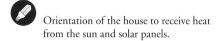 Orientation of the house to receive heat from the sun and solar panels.

 Recyclable materials for insulation and heat collection.

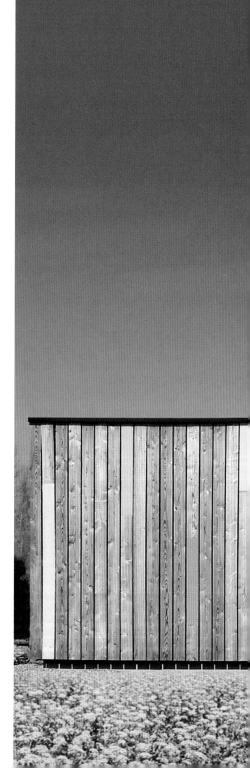

The building has been designed to optimize the absorption of solar energy. Its orientation and the materials chosen enable the heat reaching the house to be collected and retained. Locally sourced untreated wood and charcoal-colored fiber cement panels were used for exterior cladding.

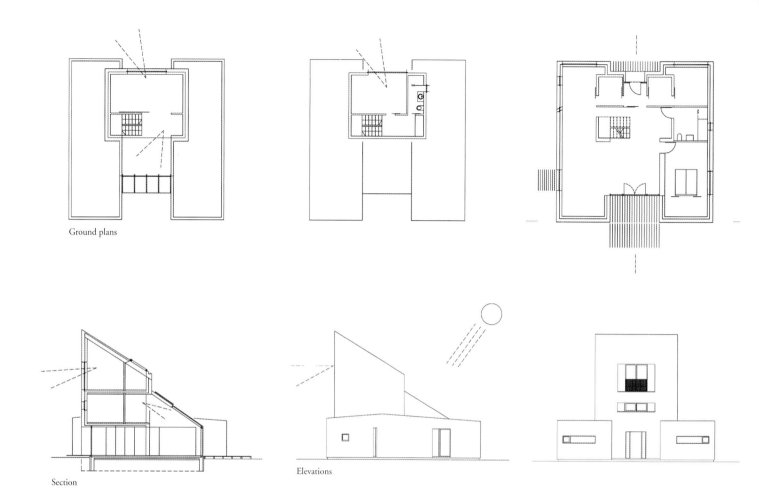

Ground plans

Section

Elevations

The plan analyzes the solar path and the orientation of the windows on each façade, showing the energetic distribution of spaces and the location of specific solar gain devices.

This house has also a partially green roof. The most obvious benefits of this passive system are the management of rainwater and reduced energy costs. In summer, they also help reduce the heat island effect.

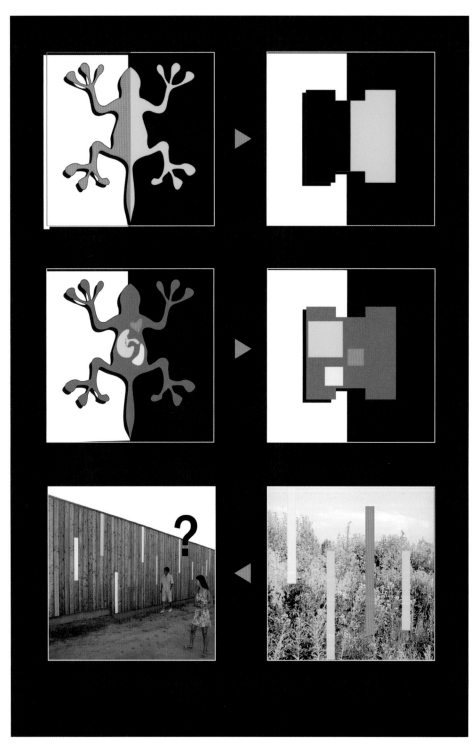

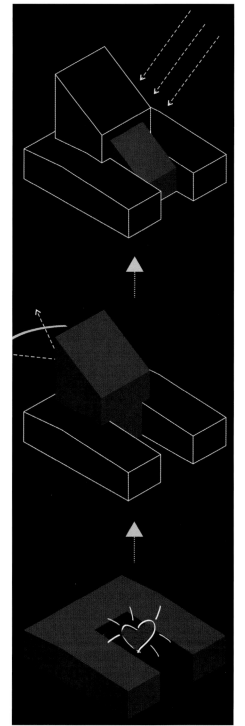

The analogy with a chameleon refers to the building's process of adaptation to the demanding climatic conditions of the local environment. The space between the two wings has been glazed over, thereby creating an enclosed collector, which receives the light from the sun's rays and heats the interior of the whole house.

Comfort and ecology join hands

Solar House | Dietrich Schwartz/GLASSX | Domat/Ems, Switzerland
© Grazia Ike-Branco

Dietrich Schwartz is an experienced architect, teacher and researcher in the field of solar energy and its use with glass façade systems. He built his first houses with energy-saving systems in the mid-nineties based on this technology, developed in Switzerland. His work is based on using the knowledge and potential culled from this field to develop products that harness the advantages and sustainability of the natural cycles of energy in a simple and cost-effective fashion. There are various types of glass, but the main purpose of this architectural glazing is to reduce the costs of the energy consumed. The system allows for the smart conversion of the solar power stored to provide appropriate air conditioning in the interior of the building, regardless of where they are located or the position of the sun during the different seasons of the year.

Solar House is one of the best examples of the work done by this company and gives us an idea of how self-sufficient a residence can be with respect to electricity consumption. It is located in Domat/Ems, a municipality in the Swiss canton of Graubünden or Grisons, in the east of the country. The hours of sunlight in this area, particularly in the winter, are not plentiful. The house nestles in the basin of the River Rhine, parallel to the valley. The east-, west- and south-facing walls have photovoltaic solar panels integrated in the large windows of the top floor. The incidence of light on the north façade is very limited, and thus storage heaters on this side of the building were not considered to be necessary. Instead, the whole façade was used to insert a picture window.

The shell is made of concrete and works like a heat accumulation system. Inside there is a bright room with exposed concrete walls, which radiate the solar thermal energy in winter just like a slow-burning stove. In this way, it is possible to make energy savings, even though it is renewable energy, which can be put to use for other purposes.

This bioclimatic residence has all mod cons, an outdoor garden, and even a pond and swimming pool for the summer months. It provides a perfect example to demonstrate that being energetically self-sufficient is not at odds with an excellent lifestyle.

 The photovoltaic solar panels fully meet the home's energy requirements.

 Exposed concrete accumulates heat from the sun and transfers it to the inside of the house.

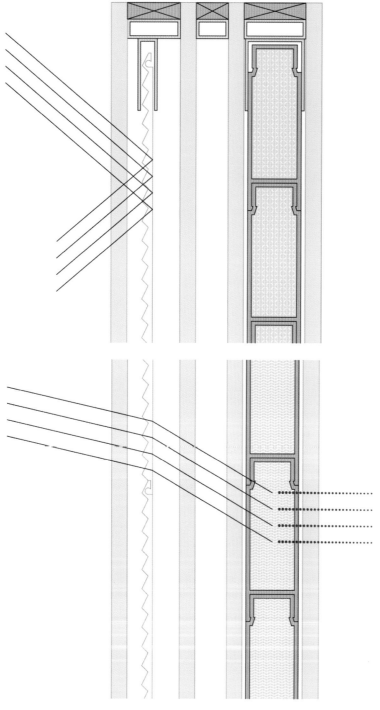

The large expanses of glass are adapted to the climate and time of day—sometimes appearing opaque and others translucent. The technology used enables the house to meet its energy requirements thanks to the exclusive use of the solar system.

Construction detail Solar radiation

In green architecture, glass is the material par excellence used to capture light. The north face —the only one without photovoltaic panels—was completed with a large window allowing natural light to enter the darker area of the house, while also making it possible to enjoy the magnificent views of the region.

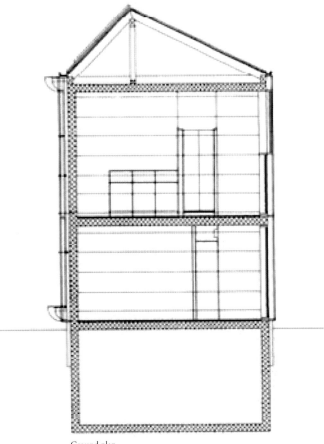

Ground plan

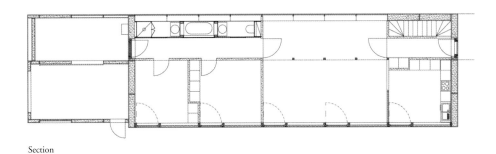

The interior rooms are spacious and full of light, and the wooden finishes add a warm touch. The exposed concrete allows heat to be accumulated and radiated to the interior of the rooms, which means that less use is made of conventional heating. The presence of glass in the walls guarantees good daylighting.

Section

Sustainability as a target for prefabricated housing

Rochedale House | Ray Kappe | Brentwood, CA, United States
© Living Homes, Gregg Segal, Valcucine

Living Homes is a company specializing in prefabricated homes that uses a steel structure allowing for a multitude of different layouts and finishes. Rochedale House, designed by Ray Kappe, is a perfect example of integration between architecture and mass construction. The architectural firm is strongly committed to the ideals of energy efficiency and environmental responsibility. Sustainable architecture has formed the focus of their professional practices for the past 20 years, and whenever possible, they work according to the LEED green building rating system.

The firm seeks to create smart buildings based on passive designer systems, such as a natural heat-monitoring system harnessing the house's orientation, thermal mass, etc. In addition, it also incorporates some active systems like photovoltaic panels. Amongst its objectives, the firm seeks to reduce the environmental impact, work with natural materials and preserve and reuse natural resources.

The house is located in Crestwood Hills, an area of contemporary residences in Los Angeles, and forms part of the housing design with large expanses of glass that blur the boundaries between the interior and exterior of the building.

Since it is a prefabricated building, the structure of the house can be erected in just three days. Apart from this new assembly system, the levels of environmental quality achieved have been recognized by the prestigious Leadership in Energy and Environmental Design (LEED) rating system. Energy savings amounting to more than 36% of the standard ratings have been made thanks to the measures implemented to achieve a reduction of the impact on the environment, air quality inside the building, savings in energy made from thermal insulation, solar protection, and renewable energy systems, for which it has earned the category of a LEED Gold rating.

The association of the manufacturer with the recycling companies will enable 76% of the materials used in the construction and assembly of the building to be reused when it reaches the end of its useful life. The house will therefore be dismantled instead of being demolished, which generates far more waste.

 Renewable, recyclable, non-toxic and locally sourced materials.

 Faucets with water-saving systems.

The prefabricated modules have been mounted on top of reinforced concrete and blocks of cement, which are good thermal mass materials. This type of modular construction has made it easier to adapt the building to the orthogonal layout of the site. The use of floor-to-ceiling windows allows for visual communication with the outside of the house.

Sketch

The modules are distributed over three levels that have been adapted to the terrain. The lowest level houses the garage and amenities. The bedrooms and bathrooms are on the middle floor. Communal areas have been incorporated upstairs on the top floor, namely, an open plan kitchen, dining area, living room, media room and the veranda.

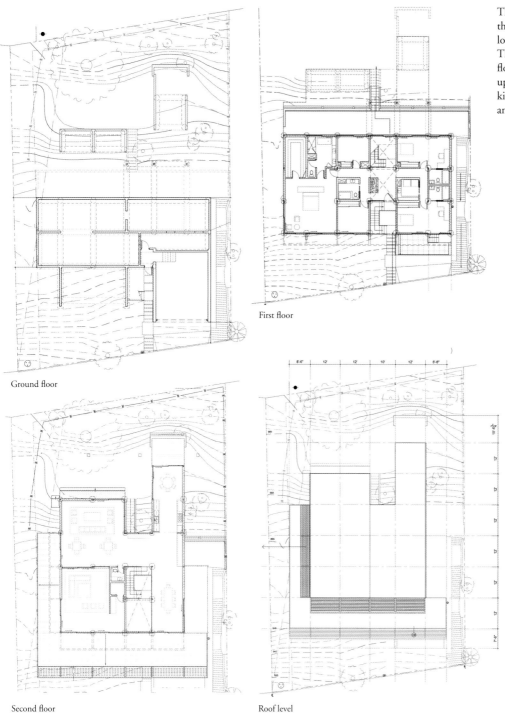

Ground floor

First floor

Second floor

Roof level

The Valcucine kitchen has also been designed according to sustainability criteria: recycled materials, wood that comes from controlled cultivation, non-toxic paints, etc. Electrical appliances, such as the washing machine, are highly efficient.

The prefabricated steel structure is visible from the communal areas. Here it is used to divide up the spaces just like screens. The steel and aluminum are recycled, the glazing is manufactured partly using reclaimed glass, and the wood comes from a certified source and has been treated with natural varnishes.

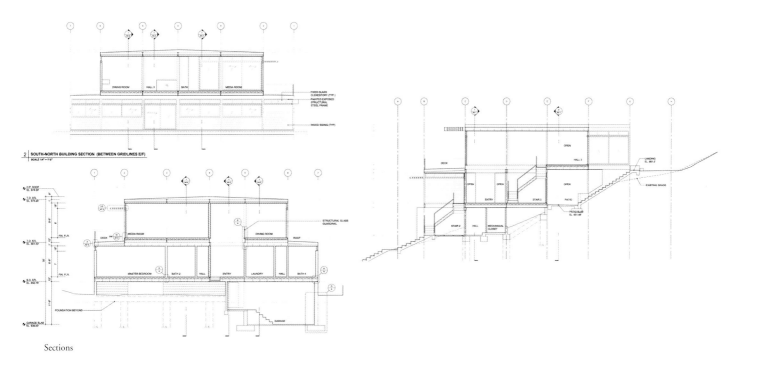

2 SOUTH-NORTH BUILDING SECTION (BETWEEN GRIDLINES E/F)
SCALE 1/4" = 1'-0"

Sections

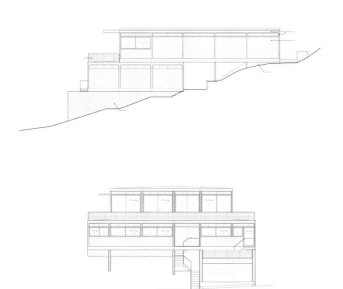

Elevations

The sections show the interior spaces of the house and the different levels. The elevations present openings and windows with low-emission glass and certified wood cladding.

Sustainability and ecology in a subtropical climate

Gully House | Shane Thompson/Bligh Voller Nield; Daniel Fox
Brisbane, QLD, Australia | © David Sandison

The house is located right next to a steep gully, in a region of dense vegetation near the Brisbane River. The property takes advantage of its privileged location to enjoy views of the lush subtropical vegetation. Due to the proximity of the river and the climate, the area is susceptible to flooding in periods of heavy rain. To adapt the house to such circumstances and leave a small footprint, it was deemed necessary to keep the surrounding trees, whose roots prevent erosion of the soil. The house was therefore erected on stilts to ensure that it was above the flood level and would permit the flow of water underneath. This building method was chosen because of the physical characteristics of the terrain and soil. The building uses a lightweight construction system that is supported on a grid of steel columns at its lowest level. The structure of the upper levels consists of a hybrid frame of steel and timber.

The house has black stained plywood cladding, which helps reduce the visual impact of the property on the surrounding environment. Other elements of the residence, including the fireplace, windows in the study and master bedroom, plus an extension to the guest room, are clad in Zincalume® steel sheets (coated in a zinc and aluminum alloy). Glass, which is also a feature of the large windows, allows the house to form part of the local environment and enables more natural light to penetrate throughout the building.

The space beneath the house (the lowest of the building's three levels) is designed to accommodate cars, to use as storage space, and to house the rainwater tanks. Access to the main floor is via a bridge across the gully, offering views to the north and east. This floor contains the spacious living room, studio, guest room, bathroom, kitchen and laundry room. The master bedroom is located on the top floor with views of the river, plus a dressing room and studio. The vegetation surrounding the property not only offers shade and assists in keeping the house cool, it also affords greater privacy.

 Rainwater collection tanks.

 The house is built on stilts and the rain water can flow underneath.

The lush, subtropical vegetation surrounding the property offers adequate insulation since it provides shade and cools the atmosphere. When the plants and trees are indigenous, specific maintenance is almost unnecessary. The plants are adapted to the environment—in this case a high rainfall zone.

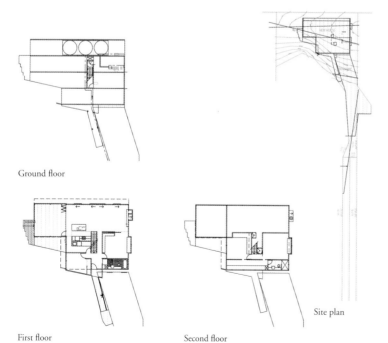

Ground floor

First floor

Second floor

Site plan

In a rainfall zone it is a good idea to install water collection tanks, either to reuse the water in cisterns and washing machines, or else for consumption following a sterilization process. It should not be forgotten, even in areas where water is plentiful, that it is a natural commodity that is in short supply. The rational use of water precludes its indiscriminate use.

Inside, the house is predominantly white and spacious. The furniture in the living room and kitchen and the wood flooring create a contemporary atmosphere. A large expanse of glass in the day area provides cross ventilation throughout the property. This huge bay window and a fan are the only elements used to cool the house.

Rationality to promote energy and comfort in Germany

Engels-Houben passive house | Rongen Architekten | Rurich, Germany
© Rongen Architekten

Passive solutions are the most common in sustainable architecture today. While active systems are developing and gradually gaining acceptance, it is quicker and easier to incorporate applications that facilitate energy savings. Despite the fact that this residence is equipped with energy-saving technology, it was built in line with standards for passive construction, which means it only consumes 14.6 kWh/m² of calorific energy per year.

The residence is located in the middle of a green area, on the outer rim of the Wurm River valley, and is surrounded by a garden and small stream. The layout of the external walls of the property has been designed with both the site and the incidence of sunlight in mind. The latter determines many of the green solutions applied within the house. The north- and east-facing walls remain relatively hidden from the sun's rays. The northern façade is usually the coldest, and the eastern face is exposed to the sunlight only in the early hours of the morning. Windows and large expanses of glass were designed for the other two façades—the west- and south-facing walls, which open out on to the garden incorporating the landscape surrounding the house. Thus, full use is made of the hours of natural daylight, thereby helping to save electricity to light the rooms.

So as to adapt the residence to the standards of a passive house, it is heated almost exclusively by means of a wood boiler, thereby avoiding the use of polluting fuels. The use of wood, known in this connection as biomass, should be assessed in terms of cost and availability of this type of fuel. If the use of wood is positive for the environment (forest clearing, surplus of farm products, etc), this should prove to be a good option. The owners also decided to incorporate a thermal system including a 750-liter tank used mainly to heat water, although it also supports the heating system. The installation enables the occupants to meet their hot water needs throughout most of the year.

 The biomass stove allows for the efficient air conditioning of the home.

 The thermal solar panels heat the water and meet the needs of the family.

Solar panels enable the sun's never-ending energy to be used. This is one of the first resources to be exploited when the aim is to apply the criteria of sustainability in residential architecture. Thermal panels enable solar power to be converted into heat.

The terrace that can be seen in the section and in the images is located over the kitchen area. The plan also shows the eaves with solar panels.

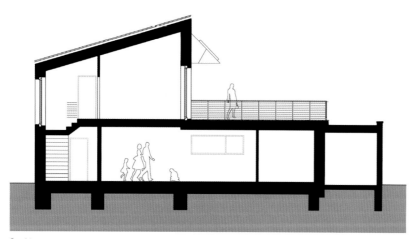

Sección

The Engels-Houben residence has a contemporary design. Its aesthetic appearance is defined by straight lines and a combination of white and gray. Sustainable architecture is not incompatible with quality finishes or a high level of comfort.

The property has a design consisting of three bedrooms, a bathroom, two toilets and a single space comprising a kitchen-cum-dining/lounge area. This area, located on the ground floor, enables the warmth from the wood stove to be shared and promotes better communication among family members.

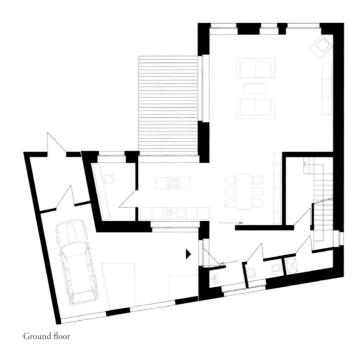

The floor plans show where the windows and glass walls have been situated. They coincide with the more inhabited spaces of the house, which need more natural light.

Ground floor

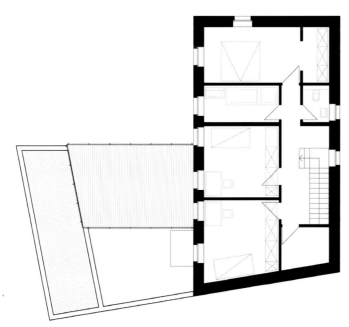

First floor

Sustainable materials for a lasting home

Modular 4 | Rockhill & Associates, Studio 804 | Lawrence, KS, United States
© Courtesy of Studio 804

This modular home with its contemporary design was created by architect Dan Rockhill and the students from the Studio 804 program, a project that involves a group of students each semester in the design and construction of a low-cost building in keeping with the assumptions of sustainable green architecture. Modular 4 is one of their latest projects.

The house, with a surface area of almost 140 m², is composed of modules that allow for the flexible allocation of space, along with a central service area that defines the interior. This enables the public sections to be set apart from the private areas, with two cubes protruding on either side of the house. Furthermore, the design of the blocks has followed some very clear guidelines as to materials and energy efficiency. The home is fully integrated in the local landscape, not only formally, but also with respect to adaptation to the climate conditions of the region. To protect the house from the cold and the winter winds, the north façade is sealed and windowless, whereas the south side of the house is completely open to the exterior and, thanks to its large windows, absorbs a considerable amount of sunlight during winter.

The property has been insulated by using cellulose in the walls, floor and ceiling. This recycled material achieves a thermal insulation that is much more efficient than that of other conventional materials. The remaining materials used in the construction of the property are a mixture of natural, certified and recycled materials. As examples, we might mention the bamboo used for the flooring, the FSC-approved wood from Brazil, or the walling systems, made of recycled aluminium. The house is also heated and cooled using passive systems, such as the skylights that ventilate the property and the white roof, which protects it from excess heat during the summer.

The house also boasts a small garden. These green spaces, with lawn, trees, plants and other items, are beneficial to the properties since they provide shade, renew the air, absorb ambient heat, etc.

The sun penetrates the house through large windows, which shed light and heat the interior.

The house has been built with natural, controlled and recycled materials.

PEFC- and FSC-certified wood guarantees that the product comes from controlled harvests, thereby preventing the indiscriminate felling of trees and forests, the deforestation of protected areas, the loss of biodiversity and socio-environmental damages to the communities living close to these exploitation areas.

The site plan shows the proximity of the house to the wooded area.

Site plan

The house has been decorated in a contemporary style. The bamboo floor provides a cozy atmosphere, accentuated by the natural light obtained thanks to the windows, particularly those in the southern façade. The distribution of the area, which follows the layout of a loft, gives the property a certain dynamic feel.

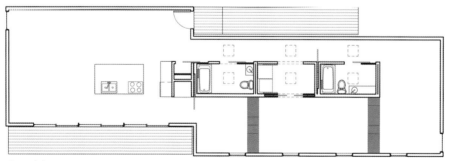

Ground plan

The use of recycled materials, such as insulation with cellulose or aluminum enclosures, allows to extend the useful life of raw materials that would otherwise become waste products that would need to be destroyed. A good approach right from the outset would not only enable to use recycled materials, but also to use materials that could be reused in future.

Walls with a life of their own

Brooks Avenue House | Bricault Design | Venice, CA, United States
© Kenji Arai; Danna Kinsky

This family home has been refurbished, a task that has involved the installation of equipment to exploit renewable energy in household chores as well as extending the living space.

Thanks to the mild climate of this area most of the year, an excellent relationship between the outside and inside of the house has been established, thus avoiding the installation of air conditioning. With this objective, the central staircase connects the ground floor with the deck on the roof terrace and it acts as a conductor pipe that draws warm air from inside the house.

Green roofs also help to regulate the internal temperature, thus three exterior walls have local, maintenance-free plant covering. Rainwater and treated graywater (from dishwasher, shower and bath) from a collection system is used for irrigation.

The home's energy needs are covered by the photovoltaic solar panels on the roof. A highly efficient boiler provides hot water that is used for direct consumption and underfloor heating.

 Photovoltaic panels that provide the home with energy.

 Rainwater and recycled water collection system for irrigation.

100

Low VOC (volatile organic compounds) paints have been used. Conventional paints contain solvents, toxic metals and volatile compounds (high VOC) which cause pollution. This happens while the paint is being applied and even when it is totally dry.

The vegetation covers three of the home's bearing
walls and also extends to the roof where plants, trees
and grass have been planted. This covering keeps
the home cool in summer and warm in winter.

The home has a system of tanks with devices
that control the release of water to prevent waste.
Reducing the consumption of this resource
means savings for the owners and savings in the
conservation of this indispensable element.

The building is raised on a steel and wood structure divided over three floors. The roof houses solar panels and the rainwater collection system.

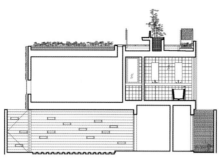

Section A

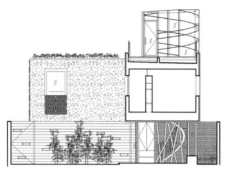

Section B

Ground level

Second level

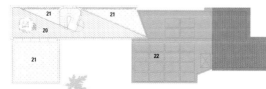

Roof level

1. Machine room
2. Laundry room
3. Store
4. Parking space
5. Main lobby
6. Kitchen
7. Living area
8. Larder
9. Store
10. Dressing room
11. Bathroom
12. Bedroom
13. Master bedroom
14. Master bedroom
15. Bedroom
16. Office
17. Bedroom
18. Bathroom
19. Roof terrace
20. Roof garden
21. Vegetable garden
22. Solar panels

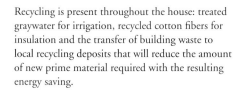

Recycling is present throughout the house: treated graywater for irrigation, recycled cotton fibers for insulation and the transfer of building waste to local recycling deposits that will reduce the amount of new prime material required with the resulting energy saving.

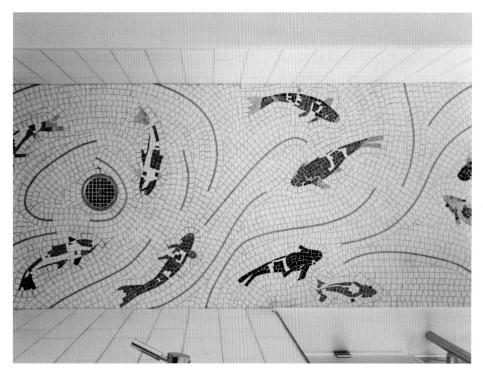

Modular housing for charming mountain refuges

Joshua Tree | Hangar Design Group | Mobile | © Hangar Design Group

Hangar Design Group is a multidisciplinary company in which architects, interior, graphic and industrial designers, all work together. They are responsible for their projects from start to finish and have made a strong showing in the field of design and manufacture of top-quality modular homes. Joshua Tree is one of the first buildings they have erected, paying constant attention to the criteria of environmental sustainability, both in the choice of materials and also in the design of the housing. Their creations are defined by their research into materials, flexibility and ecocompatibility.

Hangar practices an architecture of contrasts: Joshua Tree's outer metal structure delicately surrounds the inside of the property, which is almost entirely made of wood. Its cladding is composed of steel, zinc and titanium. It was applied in huge slats using the traditional wood tiling method. With the passage of time, the material will rust, lending the structure a more opaque appearance more in tune with snow-covered landscapes. Of note is the slope of the metallic roof, typical of Swiss chalets, since this house is particularly designed with extremely cold climates in mind.

The contrasts generated between metal and wood and the interior and exterior reflect the contrast between modernity and tradition sought by the architects. The house, while allowing for various combinations, normally has two rooms, two bathrooms with a shower in common, and a single space used as a kitchen cum dining/lounge area. The skylights in the roof and the windows allow light to penetrate the interior. It is also possible to enjoy views of the mountain peaks.

The architects and designers have placed great emphasis on sustainability. For example, the basic structure of the house is made of laminated steel, a material that can be fully recycled. There was also a concern about the environmental footprint of the building and strategies were designed to eliminate it or reduce it. On the one hand, a modular design was decided upon since, being a prefabricated home, no waste is generated during construction. On the other hand, the water, light and waste removal systems were designed so as not to leave any type of footprint on the land when the structure had to be moved to another location.

 Recycled and durable materials, laminated steel, wood.

 The building and water, light and waste removal systems do not leave any footprint on the land.

Prefabricated homes are great allies of green, sustainable architecture. Direct installation on the site considerably reduces the building time, which can be reduced from months to weeks or even days. In this way it is possible to reduce the environmental impact on the land during construction.

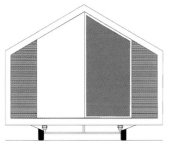

Cross elevation

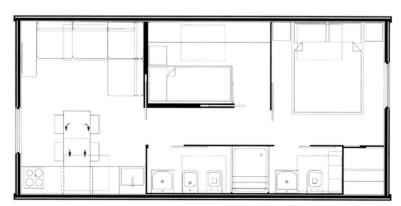

Standard ground plan

The house is designed for three or four people, or for a young family. It has two bedrooms, two bathrooms with a shower in common, and a kitchen-cum-dining/lounge area. Wood, which is used generously in the interior, offers the necessary warmth in cold, mountainous regions for which the home was designed.

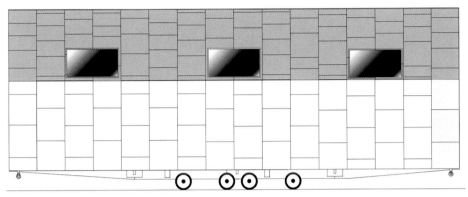

Longitudinal elevation

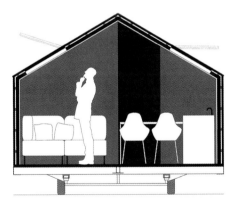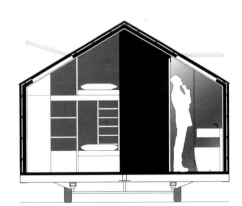

Cross sections

This construction, of only 34 m², can be moved whenever the owners like. If we add flexibility in the interior areas, these houses are perfect for use as a second home.

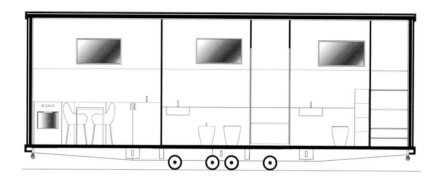

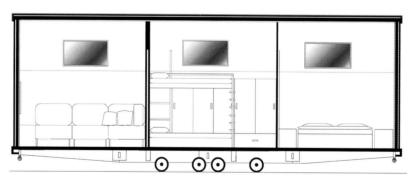

Longitudinal sections

Situadas normalmente en enclaves turísticos, estas
casas pueden conectarse a la corriente eléctrica y
al alcantarillado de la zona, aunque están previstas
otras soluciones en caso de instalarse en lugares más
aislados.

A refuge in nature with minimum intervention

Villa Långbo | Olavi Koponen | Långholmen, Finland | © Jussi Tianen

Finland is located in the northeast of Europe and nearly a quarter of the country lies to the north of the Arctic Circle, which determines its climate. Its geographic location, between the 60th and 70th parallel in the northern hemisphere and in a region with a coastal climate typical of the European continent, is the most important factor determining its meteorology. This area typically has the features of a maritime or continental climate, depending on the origin of the air currents at any given moment.

Southern Finland is washed by the Baltic Sea. The rugged southwestern coastline is repeated on the Finnish archipelago, which has over 80,000 islands, with the island of Kemiö being one of the largest. Villa Långbo is located on the western tip of the island and is exposed to the prevailing winds. The house lies on the edge of the forest and is partially visible from the sea. However, its occupants can enjoy exceptional views of the sea from any the rooms.

It is in this environment that it was decided to build a house that would have minimal effect on the ecosystem. Its peculiar location means that the house is not always accessible, it being necessary to approach it by boat in the summer months and on skis during the winter. This restricted access protects the natural environment, which is less traveled and consequently not so polluted. The idea of minimum intervention also affects the transportation of materials, which was done using horses in the wintertime, and the building techniques, since the residence was built by hand, without the use of heavy machinery. All the materials can be recycled and the wood comes from local sources.

The house cannot be inhabited all the time: some areas are located outside and are exposed to the elements. The structure is very simple and the spaces were individually designed in keeping with the lighting available. The aim was to blur the boundaries between the inside and outside of the building. A striking feature is the glazed section and a fireplace that can be used from the inside of the building and also from a covered porch.

 Recyclable materials and locally sourced wood.

 Transportation of the materials using animals, and manual construction.

The house was built on a wooden platform leveling out the ground without the need to act upon the land itself. The roof of the residence extends beyond the perimeter of the building and provides shelter from the sun and rain. The material, which is a sheet of corrugated steel, is easy to transport and recycle.

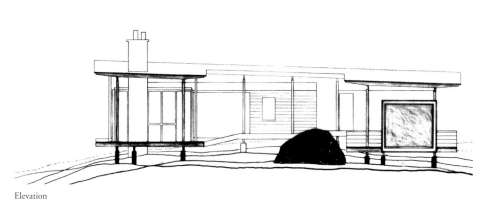

Elevation

The presence of wood integrates the house in its wooded surroundings. The different textures of wood used, such as the tiling for some of the wools and the regular slats for the floor supporting the structure, break up the monotony of the building and turn it into a highly original, cozy refuge.

The structures forming part of the house are all separate: it is necessary to step outside to move around the house. This feature responds to a wish to break up the boundaries between the in- and outside. The glazed walls of the living area and the existence of covered porches reinforce this idea.

Stunning views from French refuge

Passive House V.W. | **Franklin Azzi Architecture** | **Normandy, France**
© **Franklin Azzi Architecture, Emmanuelle Blanc**

Franklin Azzi designed Passive House for a young couple with two children. The site located on the hills near Etretat, in the Normandy region of France, benefits from breathtaking views. A ruined former hunting refuge sits on the grounds with a floor area of 18 square meters and an unusable attic space. Due to zoning regulations, new construction is prohibited and expansion of the existing structure is limited.

When the owners approached architect Franklin Azzi, the idea was clear: to build an extension to the maximum allowable. The proposal resulted in the recuperation of the derelict building and the construction of an addition to the north side of the existing building so as to remain invisible from the street and not to alter the original profile. To facilitate construction and avoid any unwanted interior partition, this homogenous addition concentrates all the stormwater drainage system and downpipes. The project also comprises the construction of two wood platforms at the second level extending off the sides of the old brick building that support demountable textile structures. These tent-like structures permit to enclose the spaces on the ground floor with camouflage canvas that blends with the surrounding landscape. The temporary spaces which may serve as bedrooms or additional common space during the summer are not prohibited by the building regulations.

Because of the remote location and hard access to the site, most part of the construction was built in the workshop, then transported and assembled on the site. The framework and the openings are all in wood and had not required any metallic parts. The wood flooring extends beyond the house walls and conceals radiant heat as well as all the water and electric piping. Per request of the owners, the house is not connected to the public water supply system and is, therefore, autonomous. This is achieved by means of a 80-meter-deep well. Another network provides the bathrooms with rainwater collected in a 2000-liter underground tank.

Future plans for the house include the installation of photovoltaic panels on the north side that will make the house self-sufficient.

Overall, this simple but enthralling project responds to strict environmental requirements, especially when it comes to collecting rain water, the use of solar and geothermal energy.

 Hot water and heating are obtained by means of a combination of hybrid solar and geothermal systems.

 Rainwater harvesting to meet non-potable water needs. A 80-meter-deep well supplies drinking water.

Restricted by the local development plan, the expansion of the existing house consists of a two-storey wood-clad gabled volume that adds 6,5 square meters to the existing building and two wood terraces on the second floor, each producing a sheltered area of 17 square meters on the ground floor.

The symmetrical camouflage canvas tents provide a playful and temporary solution to expand the floor area without sacrificing comfort. A second phase of the project includes the construction of a concrete bunker buried in the hillside below the house with a fully glazed wall facing south and the views of the valley.

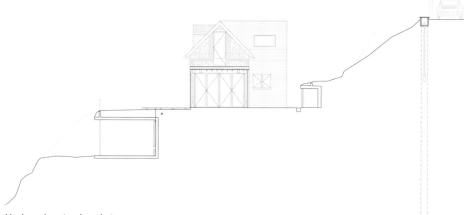

North-south section through site

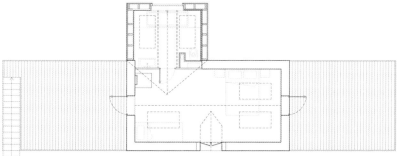

Second floor plan

To optimise the surface of the liveable space, all equipment such as the boiler, the water pump, the electrical meter and the garden shed are located in a partially buried space dug in the hillside facing the kitchen. East of the house, an existing shed was reused to function as a fully equipped bathroom.

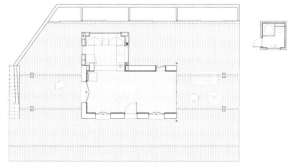

Ground floor plan

The existing building houses the living room on the ground floor and a large bedroom on the upper floor while the kitchen and the bathroom are stacked in the new wood-clad extension to facilitate the running of water pipes.

Sustainable family beach home

Seadrift Residence | CCS Architecture | Stinson Beach, CA, United States
© Matthew Millman

The architects at CCS Architecture designed this spacious property as a second home for a 3-generation family. The house is located in Seadrift, a gated vacation community that originated in the 1950s at the tip of Stinson Beach.

The design of this family house follows a rigorous sustainability and energy efficiency program. All the home's systems—electricity, hot water, HVAC, and radiant heating—are electric-based, and powered by the photovoltaic panels on the roof. With the exception of the cooking range, which uses natural gas drawn from a propane gas tank, the whole house has zero energy consumption, thereby avoiding the emission of large quantities of CO_2. The residence is a perfect example of how opting for energy efficiency can be compatible with comfort and convenience. Furthermore, in anticipation of rising sea levels, and in compliance with the regulations in force in the region, the building has been erected about one meter above street level.

The plan of the house has been designed to accommodate a large number of people at different times of the year. It is appropriate for having both family and friends round, and therefore the daytime area is of particular importance. The living area opens directly on to two courtyards outside. The window walls on both sides provide uninterrupted views of the hills and water. The courtyard on the south side is more intimate whereas the one on the north of the building leads to a small jetty in the lagoon and also contains a small pool.

The house, which only has one story, is divided into two main wings: the social enclave, which includes the living room, kitchen, dining area and a multipurpose space; and the one containing the three double bedrooms, separated by the two bathrooms. The outer decks or courtyards extend the surface area of the home, particularly in the summer, when more time is spent in the living and dining areas outside in the open air. The fire orb in the living room can be rotated to direct its warmth to the interior of the house or towards the courtyard.

 Photovoltaic panels, which provide all the electricity in the house.

 Rainwater collection system for watering the garden.

The whole house is covered with wood siding. The product for treating this material for use on external structures does not contain chromium or arsenic. The property was erected approximately one meter above street level, in compliance with current building codes, to enable it to withstand a potential rise in the water level.

The terrain, surrounding gardens and plants in the courtyards are watered using a drip irrigation system, which is far more efficient. A system has been installed to filter and store rainwater. The storage of water helps with the watering of plants and the tank collects water and reduces the risk of flooding.

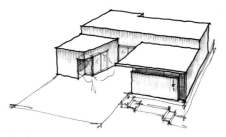

Sketch

The use of solar energy does not require the construction of special spaces to accommodate equipment and gives the architect absolute freedom in the design.

To control water consumption, the architects installed flow reducers on all faucets and showerheads. These systems, which normally mix air and water, obtain up to 50% savings in water. Regulators were also installed to control the consumption of light to rationalize the use of electricity.

The plan shows which are brick walls and which have glass openings. This allows us to see the possible combinations to generate air currents.

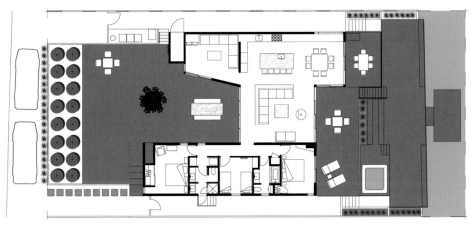

Ground plan

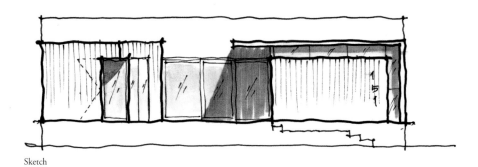

Sketch

Fabulous views from a rocky enclave

Mataja Residence | Belzberg Architects | Santa Monica, CA, United States
© Tim Street-Porter

In the mountains of Santa Monica the weather is hot and sunny, with very little rain throughout the entire summer. The dry climate means that the place is a high-risk zone for forest fires. In winter the climate is cold and more humid, but is not known for snow, save in exceptional cases, as the highest peak is only 950 m.

Mataja Residence is located high up on a rocky enclave, which has had a profound influence on the design of the building and the approach to the house program. The model of a typical California courtyard house was taken as the starting point for the project, albeit with modifications due to the large rock formations on the site, which have been integrated inside the house in a very original and skilful fashion. Furthermore, the rocks protect the courtyard from strong winds and offer privacy. The courtyard provides a focal spot from which the rest of the house radiates, occupying an area of 418 m².

The house, divided into various spaces, has a fragmented appearance, lending it a highly dynamic air in spite of its large dimensions. The type of coverings used accentuate this sensation. The metal roofs not only make the construction more resilient but also give it a modern touch. A studied 32-degree incline allows it to capture the maximum amount of light and facilitates the harvesting of rainwater. There are some water tanks for storing rainwater for irrigation purposes. Together with the choice of local vegetation that does not require much water, these elements have generated outdoor areas that are self-sufficient.

The home's thermal insulation is achieved thanks to the concrete walls and flooring made from slabs of the same material. The concrete retains the heat from the sun, because of its physical properties, and passes it to the interior of the house at night, thereby making savings on heating. Due to the structure of the building, it was possible to use glass in 67% of the interior, making much better use of the natural light. The large expanses of glass offer impressive views of the landscape surrounding the house.

 Capture of the maximum amount of light thanks to the slope of the roof.

 Harvesting of rainwater to irrigate the site.

The openings in the façades and the presence of the courtyard enable the house to be properly ventilated. The vegetation of the outdoor areas is not exaggerated—quite the contrary, since various types of cacti can be seen. Finding the appropriate vegetation for the natural environment respects the typical ecosystems of the area.

Site plan

The sections show the exact angle of the roofs which collect rainwater. A courtyard has been formed from the rocks of the site so as not to ruin the environment, this can also be seen in one of the sections.

Sections

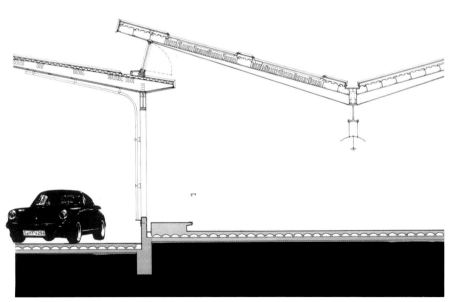

Construction detail

Apart from the residence itself, the house also has an addition with a garage, a guest house and a spa. The aim was to build a second home where the owners could escape from their urban lifestyle and take refuge in a more isolated setting to entertain family and friends.

Thanks to an exhaustive study of the terrain and setting, a residence emerged with its different areas nestling between the rock formations, as an integrated natural prolongation of the landscape. The huge rocks have been integrated very skilfully inside the property, with hardly any disruption of the landscape.

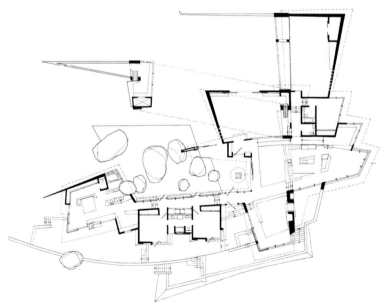

Ground plan

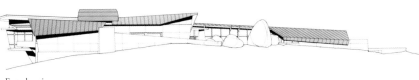

East elevation

North elevation

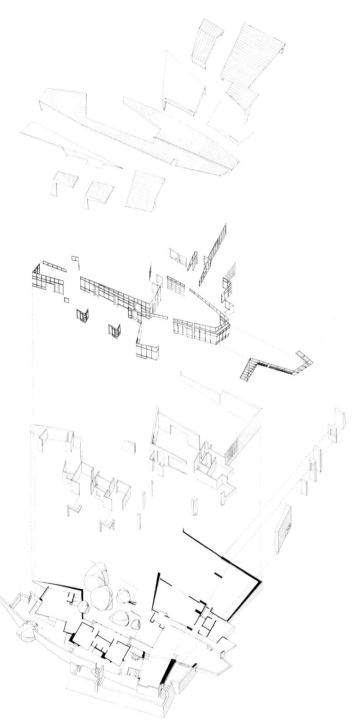

Exploded view

Architectural sensitivity for an environmentally friendly house

Joanopolis House | Una Arquitetos | Joanopolis, Brazil | © Bebete Viégas

This property is located on the border between the states of São Paulo and Minas Gerais, more specifically at the foot of the Maniqueira Mountains, at an altitude of about 1000 meters. The house is part of a housing development on the shores of Lake Piracaia. The buildings of this type of residential complex are usually grouped together, spreading out to occupy the maximum amount of space on the site. In this case, however, the structure of the house has been adapted to the slope of the land and a large space has been generated around it, keeping it some distance from all the other houses.

Various techniques were used to minimize the impact on the environment. For example, soil excavated from the ground was used to refill other areas wherever it was needed thereby avoiding any excessive shifting of earth. The retaining walls were built with stone collected from the same site and traditional techniques were also employed to erect them.

There are three courtyards that link up directly with the areas outside the house, thereby preventing the rooms from being cut off from the surrounding environment. The stone walls protect the house from the wind giving it privacy. At the other end of the house, out in the open, is a small swimming pool, in an area with views of the sea. The green roof represents another ecological strategy for obtaining good insulation in the house, since it ensures an effective thermal inertia in a place where there are significant thermal differences during the day and night. It provides a way of controlling the indoor temperature without the need for air-conditioning.

There is a soaring white tower which seems to emerge from the roof that links up all the infrastructures of the various amenities: kitchen, heaters, piping and the water tank for rain-water collection. Concrete has been used for the structure of the building and the walls and finishes have followed one of the premises of the house design: the savings in cost. The glass façades provide cross ventilation and good lighting in the house. So as not to incur greater expense, the qualities inherent in the materials were respected and unnecessary finishes were avoided.

 Elements for natural insulation: a green roof, gravel in the courtyard, etc.

 Rainwater collection.

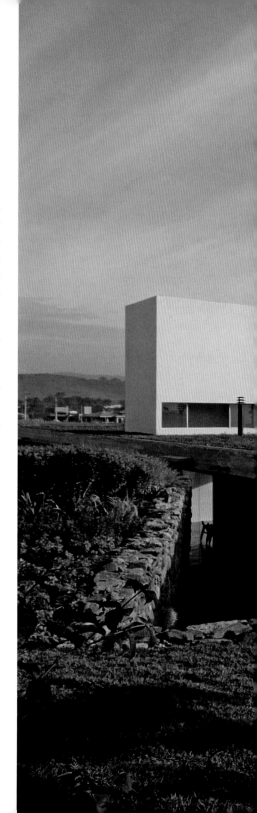

The gardens have been planted with indigenous vegetation, adapted to the rainfall of the area: peroba rosa, jequitiba, embauba, aroeira, trauma, Brazilian walnut (ipê), grumichava, jacaranda caroba, acacia and jacaranda. Fruit trees were also planted outdoors close to the house: Jabuticaba trees, orange trees, lemon trees, etc.

The cross sections show the uneveness of the land and indicate different elements of the home: chimney, swimming pool, walls surrounding the back patio, etc.

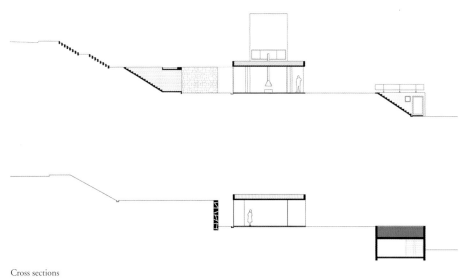

Cross sections

To minimize the impact on the land, the house
follows the contours and sloping terrain of the site
and its orientation has been designed to take full
advantage of the sea views. The green roof and the
gravel increase the thermal mass of the house and
facilitate a more natural and efficient form of air-
conditioning.

On the main façade the glazed surface was shifted to enable the roof overhang to create a porch to protect the interior from the sun. On the rear façade it was not necessary to make the same division since the stone walls shelter the house from the wind. The gravel in the courtyards absorbs the heat during the day, which is then radiated to the interior of the house during the night.

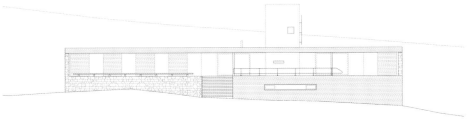

Longitudinal elevation

The site plan shows the distance and the slope from the house to the water. The plans in the next page shows the openings of the house and the back walls that protect the house from the wind. These elements regulate ventilation inside the house.

Site plan

158

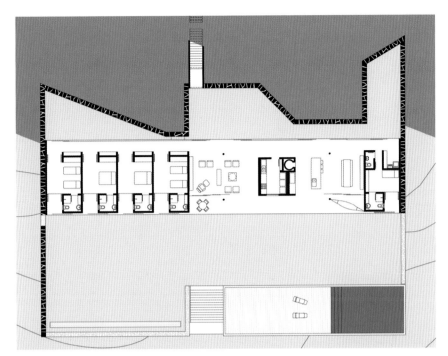

Top floor

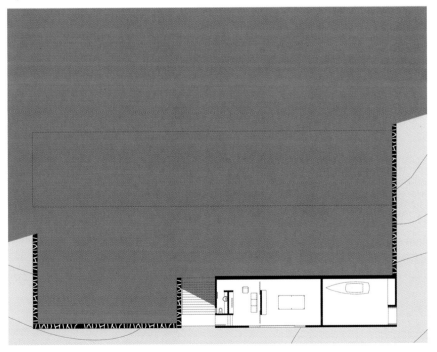

Lower floor

Holiday home in paradise

Lake House | Casey Brown Architecture | Port Stephens, NSW, Australia
© Rob Brown

The area of Port Stephens is known as the blue water paradise because of the beauty of its water and its natural environment. It is located about 160 km northeast of Sydney, on Australia's eastern seaboard, and has a population of approximately 50,000 inhabitants. A natural beauty spot, it is divided into three small regions—the Tomaree peninsula, the Tilligerry peninsula and the Golden Bight area surrounding the bay.

The climate is warm all year round, with no winter frosts. The cool breezes keep the temperature mild in the summer. During this season it is a great escape for Sydneysiders who flock to its beaches, which are much quieter than anything Sydney has to offer. In the spring, the area comes alive with indigenous wild flowers, and from May to July it is possible to see the migrating birds fly through these lands.

It is in this idyllic spot that a young family decided to build a summer home on the edge of a lake, next to a eucalyptus forest. The environmental concerns of the owners led to the inclusion of passive energy-saving systems in the design.

The house is built on a platform to guard against flooding and avoid causing a significant impact on the local terrain. This surface provided the base on which the seven pavilions forming the residence were erected. These structures are articulated around a central passageway crowned by a huge open fireplace. Access to the raised area is via two ramps, one leading to the lake and the other to the bush. The rooms in the house are distributed between the pavilions, and the disposition of each one is dictated by the climate and the use made of the room. Their position with respect to the lake, air currents and need to regulate the temperature also has an influence. The pavilions also collect rainwater in four underground tanks, some of which are reserved for firefighting purposes. Another of the environmental measures worthy of mention is the treatment of waste using a recycling system. Sliding walls and louvers in all the pavilions open the house to the sea breeze and the two-sided fire place warms the main pavilion.

 Passive systems for heating and cooling the pavilions.

 Water collection system for consumption and firefighting purposes.

The upper part of the walls of the pavilions is clad with a translucent material that allows natural light to enter and avoids greater energy consumption for lighting purposes. That material is fire-resistant, and was chosen because the house is located in a high risk zone for forest fires.

The interior of the pavilions is covered in wood, which is an extremely recyclable material. Sliding walls and louvers keep the spaces ventilated, by using the orientation of the house and the presence of sea breezes. This type of ventilation saves energy, particularly during the hottest months of the year.

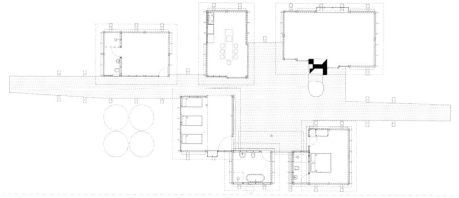

Ground plan of the main pavilions

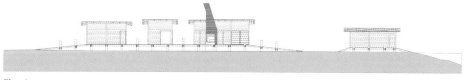

Elevation

The idea of building discrete independent pavilions for each unit arose from the intention not to predominate the landscape. The separation between the pavilions is to make better use of daylight and breezes.

The distribution of the house into separate
pavilions enhances the property's continuity with
the surrounding countryside, since a large part of
the occupants' everyday life is spent outdoors and
so there is more contact with nature. The ramp
leading to the lake also goes to a seventh pavilion,
located on its shores.

Organic shapes and sustainable architecture

Domespace | Patrick Marsilli | Quimper, France | © Benjamin Thoby

This strange circular house is located in the heart of Brittany, France. It provides a model for a large number of similar homes, the majority of which are located in France, where their creator, architect Patrick Marsilli, first started to design in 1988. His ideas and actions are considered to be the precursors of domotics (home automation) and sustainable architecture. The origin of this shape drew its inspiration from the religious architecture of the cathedrals and the shape of the pyramids.

The peculiarity of this house is its ability to rotate in the direction of the sun or in the opposite direction. This feature was first included in the most modern automated homes. The dome can be rotated by hand or by remote control. The house has a maximum angle of rotation of 320 degrees and the software controlling the automatic movement also adjusts the angle and speed of rotation.

Another important feature in the design is its resistance to hurricanes and earthquakes. While it may indeed be true that these two natural phenomena are not common in France, the possibility of building this type of home on other parts of the planet makes this feature quite attractive.

Maximizing use of natural daylight and the large central space inside the dome are two of the advantages of living in this type of building. The shape follows the standards of geobiology (the science that studies the interactions between life and the Earth's physicochemical atmosphere). The house is free of electromagnetic disturbances, water flows and radioactivity.

The organic geometry of the residence is adapted to the natural environment: it is similar to the typical structure of an igloo or the shell of an animal. Its structure, which is mostly made of wood harvested from FSC-certified forests, generates the floor, a central column and some arches that run from the end of the column down to the ground. The roof is made of cedar wood, a species that is highly resistant to rotting. Cork was used to make the residence adequately heat- and soundproof, while at the same time obtaining good acoustics inside the house.

 Passive solar energy and cross ventilation.

 Organic shape and geobiological design. Automatic orientation towards the sun.

The source of the wood is regulated. FSC certification ensures that it is not sourced from companies that destroy protected forestland or from unregulated areas. Apart from preventing deforestation, thanks to these certificates it is possible to verify the physical characteristics of the wood and guarantee its quality.

The house is heated and cooled by orienting the house towards the sun. In addition, a central fire place provides extra heat during the coldest months of the year. In summer, the efficient cooling of the interior of the building is achieved by means of cross ventilation. These passive systems avoid wasting energy and ensure the comfort of the people in the house.

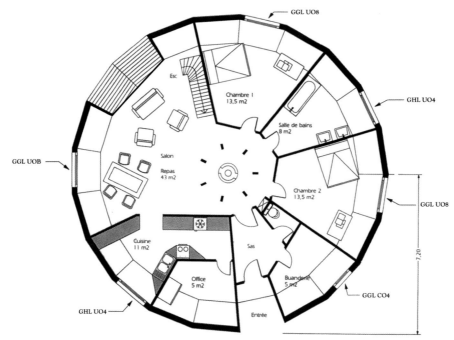

GGL UO8

GHL UO4

GGL UOB

GGL UO8

7,20

GGL CO4

GHL UO4

Esc

Chambre I
13,5 m2

Salle de bains
8 m2

Salon

Repas
43 m2

Chambre 2
13,5 m2

Cuisine
11 m2

Sas

Office
5 m2

Buanderie
5 m2

Entrée

Ground floor interior distribution model

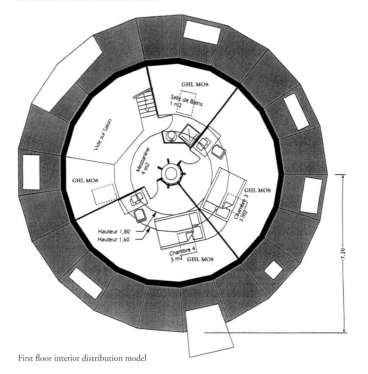

GHL MO8

Salle de Bains
1 m2

Vide sur Salon

GHL MO8

Mezzanine
4 m2

GHL MO8

Chambre 3
3 m2

Hauteur 1,80
Hauteur 1,60

Chambre 4
3 m2

GHL MO8

7,20

First floor interior distribution model

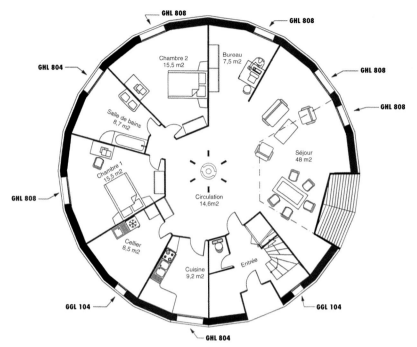

Ground floor interior distribution model

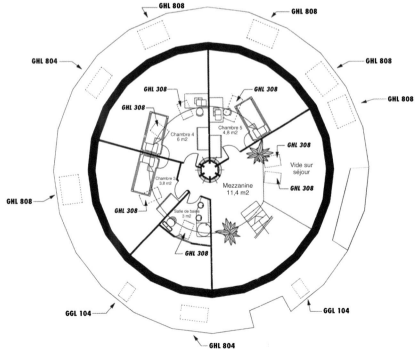

First floor interior distribution model

Patrick Marsilli's design supports a number of different options, ranging from the surface of the residence (various sizes are available ranging from 44 to 200 m^2) to the distribution of the spaces inside the building. The interior of this house is built almost entirely from wood: walls, furniture, floor, etc.

A green cabin next to the ocean

House in Buchupureo | Álvaro Ramírez, Clarisa Elton/Ramírez Moletto
Buchupureo, Chile | © Álvaro Ramírez, Clarisa Elton

The Bio Bio region, where the town of Buchupureo is located, marks the transition between the dry temperate climates of the central zone of Chile, and the wet temperate climates found immediately south of the river Bio Bio. This region forms the boundary of the last stage of the Mediterranean climate.

The house is located in an area that enjoys a fairly benign climate, with few extremes. This means that it does not require robust insulation, and passive systems can be used for efficient air conditioning. The construction, which is of reduced dimensions, has been adapted to suit the local environment and the rocky terrain forming the surface available for the site, on the edge of a cliff. The house is perched on the abrupt slope on top of stilts that reduce the impact of the building on the ground and at the same time allow for the natural draining of rainwater. The rooms are distributed in two blocks. One of these contains the living room, kitchen and dining room and the other, the bedroom and bathroom. Between the two modules there is a wooden platform offering an open space facing the sea, which is protected from the breeze blowing in off the ocean. This open porch offers an additional space that acts as an outdoor dining area or sun lounge and links up the public and private areas of the house.

The whole façade faces the sea and consists almost entirely of glass, allowing the hours of sunlight to be enjoyed to the full, along with the views. The space between the two modules comprising the house allows the property to be suitably ventilated, thus protecting it from exposure to extreme heat during the summer. The blinds installed inside the house and the trees nearby provide the necessary shade. The materials used—pine and slate for the roof—have become deeply rooted in the local architecture. The use of locally sourced timber, which enables savings to be made in terms of cost and CO_2 emissions, and the design of the house (which is respectful of the orthogonal layout of the site) are two of the most striking elements of this small dwelling.

 The orientation of the house maximizes the use of sunlight and favors cross ventilation.

 The stilts respect the land and permit the drainage of rainwater.

176

The house provides a small refuge where you can switch off and enjoy nature and the ocean. The ground on which the property is located is extremely rocky and uneven, but the architects made it a good site by using stilts for support. This system also avoids excessive impact on the land.

The site plan shows how the construction, similar to a cabin, adapts to the environment and the unique terrain on which it rests, at the edge of a cliff.

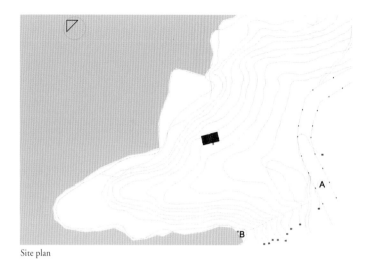

Site plan

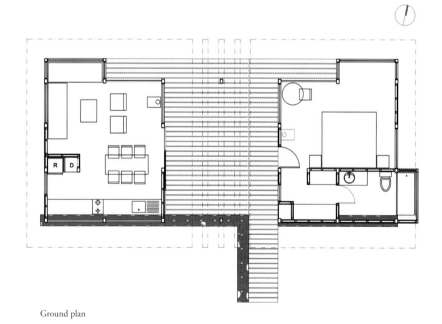

Ground plan

The open porch has a dual function. On the one hand it helps to ventilate the house, particularly in summer, making use of air conditioning or any other type of cooling system unnecessary. Furthermore, it constitutes another outdoor area, functioning as a link between the public and private sections of the house.

The use of slate and timber as building materials creates a warm, cozy atmosphere. These materials are both ideal for a refuge of this type, with its simple design and reduced dimensions. Adjusting the dimensions of a house for the specific use it is to be given saves on the cost of materials and maintenance.

Alpine chalet close to the Finnish lakes

Villa Nuotta | Tuomo Siitonen Architects | Kerimäki, Finland | © Rauno Träskelin, Mikko Auerniitty

This magnificent retreat was built on the northern tip of the island of Herttuansaari, south east of Kerimäki and between lakes Puruvesi and Saimaa, the largest lake in Finland. The ground is rocky and sparse in vegetation—quite barren. Its main feature is its steeply sloping terrain, covered only by pine trees. Throughout most of the eastern side, the hills drop sharply down to the coastline and the place is a mixture of shrubs and scrubland. To the north, the stones are covered with moss and lichen. The views that can be enjoyed form part of Natura 2000, a network of natural areas protected by the European Union, which also includes parts of Latvia and Spain.

The project entailed designing a holiday home that would only be occupied at certain times of the year by a couple of adults or a small group of friends. The property, moreover, has all the amenities for larger periods of time to be spent there—and even for work purposes.

Albeit due to the natural setting where the house is located or because of an environmental commitment, the architects have used local sandstone and certified timber throughout the entire construction of the house. The use of this type of material significantly reduces the impact on the environment. The timber is controlled by the Finnish Forest Certification Council, a body comprising one of the world's largest wood producers, which ensures sustainable practices are used in harvesting this important resource. The wood used has been subjected to two different treatments, but both are equally respectful of the environment: colorless varnish and black varnish made from vegetable oils.

Furthermore, the architects decided to respect the terrain where the house was to be erected as much as possible, and thus opted to adapt the ground plan to the site rather than adjusting the nature of the ground itself. To do so, the spaces were fragmented and spread around the site. A patio oriented towards the sun offers some splendid views of the lake.

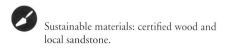 Sustainable materials: certified wood and local sandstone.

 Respect for the local terrain: the ground plan of the house has been adapted to the orthogonal layout.

The use of certified, natural materials in building the property ensures sustainable construction. For example, the interior of the house was built entirely with local timber. Sandstone, also sourced locally, was used for the outside of the house.

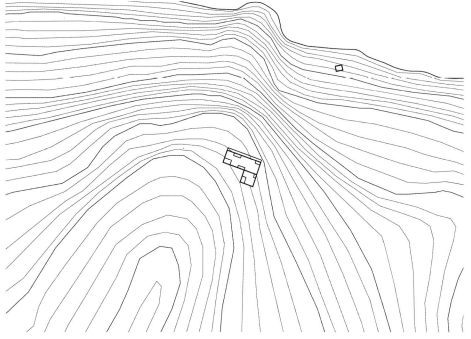

Site plan

As can be seen in the site plan, the house is perched on top of a mountain, near the edge of a cliff. This location, along with the skill of the architects, made it possible to design a house affording views of the lake. The glass walls make it easier to admire the landscape in winter.

The inside of the property has all the amenities needed to spend longer periods of time at the house. The top floor—with its glazed walls designed to ensure this part of the house receives more sunlight—is the area most frequented by the owners: the adjoining space between the living room, kitchen and dining room, the bedrooms and a sun lounge. Downstairs the presence of a sauna takes pride of place.

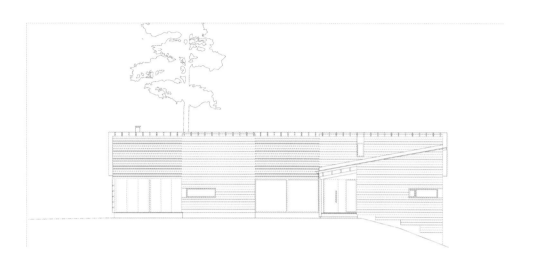

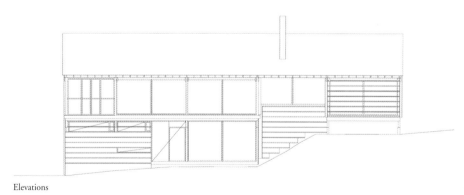

Elevations

The front and rear elevations show the characteristic features of the façades: on the front façade glass walls facing the lake stand out, which contribute to solar gain depending on the time of year. The rear façade, clad in wood, is not very permeable for the outdoors.

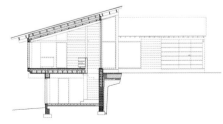

Section

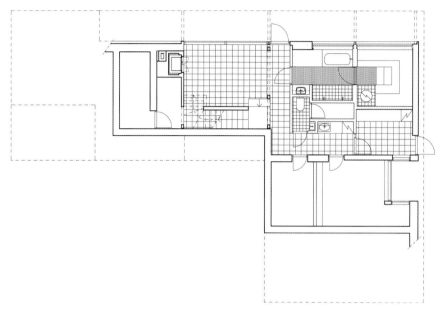

Ground floor

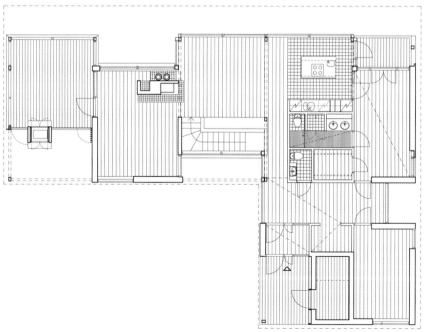

First floor

The interior of the house has everything you need to spend long periods of time. The glass façade and cantilevered upper floor promote solar gain. This floor is the most occupied, as it includes a living area, kitchen and dining room in addition to the bedrooms and a lounge. The sauna, an essential element of Finnish homes, is located on the ground floor.

Simplicity and harmony for a house in equilibrium with its surroundings

Lake House | UCArchitect | Marmora, Canada | © UCArchitect

This small retreat was built in Peterborough, near the town of Marmora, close to the US border. The house is almost impossible to see from the lake but there is a perfect view of it from the access road.

The house has spectacular views of the lake. It is surrounded by dense vegetation, with the predominance of trees such as pines, cedars and other local species.

With respect to the structure of the house one of the main features is a wooden L-shaped screen, which marks the entrance and generates a wrap-around deck on the outside of the building. Furthermore, a very important link has been set up between the in- and outside areas of the house. Apart from having scenic views, this relationship is accentuated by the existence of skylights, roof overhangs and large openings instead of traditional windows. In winter these afford views of the exterior without it being necessary to go outside and brave the weather, and in the summertime the openings blur the boundaries between the inside of the house and the vegetation outdoors. The house is accessed by means of a stairway, built on piles so as to avoid the need to alter the terrain any more than really necessary.

Various strategies are used to enhance sustainability, particularly of a passive nature, which make it posible to ensure that the house is well insulated by harnessing nature itself, i.e. wind, sun, etc. One of the most significant is the use of cross ventilation. Studying the air currents and openings means that adequate ventilation can be achieved that keeps the house cool in summer without the need to install HVAC equipment. Reinforcing the insulation is another of the strengths of this building. The cold winter climate means that there is a greater difference in temperature between the in- and outside of the house, which makes it necessary to prevent the heat from escaping. The use of passive solar power is another system used in the house. The openings and skylights provide good lighting. The flooring in the house is made of concrete, a material that acts to reinforce the radiation from the ground, since it possesses a large amount of thermal inertia—it stores heat during the hours of sunlight then radiates it to the interior of the house.

 Passive systems: reinforced insulation, cross ventilation.

 Respect for the terrain and the environment, and preservation of the local vegetation.

The project perpetuates the identity and character of the surroundings, in which nature plays a major role, on an appropriate scale composed of small spaces. Purpose-specific dimensions for the use to be given to the residence make it easier to insulate, light and keep clean.

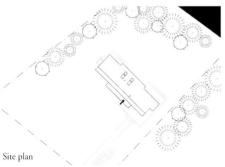

Site plan

The interior of the property has a contemporary style. There is a predominance of grays and blacks, particularly in the floor, bathroom and kitchen furniture. The white walls and pale wooden elements create an interesting contrast that lends the decoration a certain vitality and air of modernity.

The side elevation illustrates the particular inclination of the roof, which adapts to the path of the sun to control excessive radiation and achieve optimum thermal comfort and lighting inside.

Cross elevation

The layout of the interior has been obtained by arranging the rooms around the central core formed by the kitchen and bathroom. Three sliding doors that can be combined in a number of ways are the only partitions on this floor. A wood stove completes the day area.

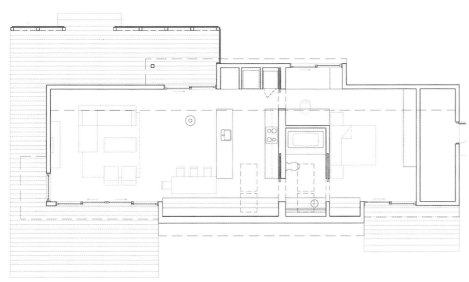

Ground plan

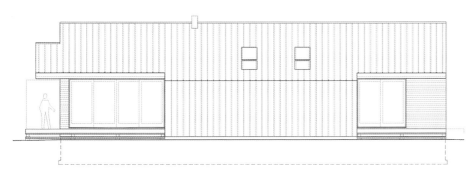

Longitudinal elevation

A retreat to escape from daily routine

Lovely Ladies Weekender | Marc Dixon | Walkerville, VIC, Australia
© Lucas Dawson

Three friends decided to build themselves a weekend home on the coast. Their projects were based on cost containment and sustainability criteria. Initially, the owners did not think of hiring an architect, assuming such services would not meet the constraints of their budget. Finally, they contacted the architect Marc Dixon, who accepted the challenge offered by his clients.

One of the first things required was to leave as small a footprint as possible on the site, which is located in an area reserved for holiday homes, with very limited infrastructure. The town has no definable centre, there are virtually no shops, and there is no running water or sewage. 80% of the site is covered with remnant bushland—indigenous shrubs and plants. The property is some 800 meters from the beach with its boundary abutting on to rural land.

Design considerations have mainly concerned issues of an environmental nature. It was necessary to increase the density of the vegetation, which also needed to provide more privacy for the lowest level of the house.

To achieve a low-cost home, construction techniques needed to be simple. The structure had to be lightweight and it was necessary to look at simple solutions for the cladding. The availability of local building materials was also considered. In addition, tanks were installed to harvest rainwater. The material used to clad the house and tanks consists of prefabricated corrugated sheet metal, which is widely used in the area.

The ground plan of the house is long and narrow—with a maximum width of 4 m—and needs to fit into the space available. This shape maximizes the hours of daylight and facilitates cross ventilation, two of the most important passive systems used for obtaining adequate insulation inside a home. The top floor fully exploits the potential views of the surroundings. The longitudinal axis of the building is perpendicular to the north, thereby achieving excellent orientation to the sun in the southern hemisphere.

Cross ventilation and orientation of the house to take advantage of daylight hours.

Rainwater collection tanks.

The house is located on a wooden structure that also forms the patio on the ground floor. A balcony has been constructed on the floor above, with views that can be enjoyed from the dining room. The tanks used to store the rainwater and the cladding on the house are made of Colorbond®, steel sheeting in a variety of colors.

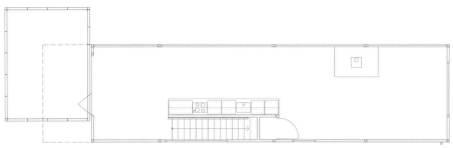

First floor

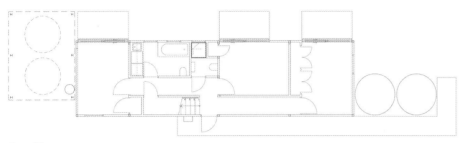

Ground floor

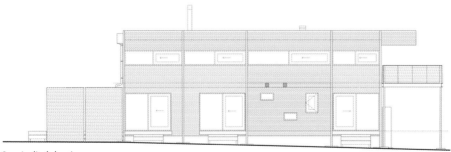

Longitudinal elevation

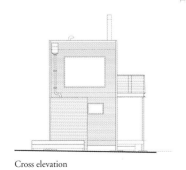

Cross elevation

The house is of minimal dimensions. The three bedrooms, each of which has direct access to a secluded deck, are small. Two bathrooms have also been installed on this floor. The communal areas—kitchen, dining room and living room plus fire place—are located on the upper floor. The small openings on the southern side facilitate cross ventilation.

Cross elevation

The narrow, orthogonal layout was designed to minimize costs and reduce the waste normally generated in the building process. Another cost-saving measure is the use of prefabricated materials. For example, for the finishes on the external cladding, prefabricated metallic edges were used.

Natural materials for a forest setting

House on Lake Rupanco | Beals Arquitectos | Lake Rupanco, Chile
© Alejandro Beals

This house is located on the shores of Lake Rupanco in southern Chile. The region has a temperate climate with temperatures ranging from 9 ºC to 18 ºC. On the site there is a slope of almost 36 meters and views of the lake. The orientation of the house has been designed to take account of this incline in the terrain to make the house overlook the lake and take full advantage of the views. The main façade is glazed to create visual continuity with the lake and the surroundings. The rear wall is lost in a forest of native shrubs, myrtle and laurel.

There are a certain number of similarities between this residence and Reutter house designed by Matthias Klotz. The entrance is reached via a wooden walkway located on the top floor at the end of a path winding through the trees before finally arriving at the house. The density of the vegetation around the house offers a greater feeling of privacy.

The lack of good access to the site for machinery and the transportation of materials made the architects opt to use wood and hire local workers that were used to working with this material. Thus, the structure is made of pine. The outer walls are also pine, on this occasion impregnated and coated with carbolineum, following a technique used in rural areas in southern Chile to seal the wood and protect it from the persistent rain prevalent in the region. Carbolineum is a liquid substance obtained through the distillation of coal tar. The interior facings and flooring are of untreated mañío and ulmu wood respectively. The structure of the residence was based on the shapes and textures used in other buildings in the area, such as stables and granaries.

The layout of the house program gives priority to the master bedroom and the day area, these being the two spaces with the best views. The double height of the day area enables more light to penetrate and makes the cross ventilation more efficient. On the eastern side, a garden separates the two bathrooms, ensuring that natural light reaches both rooms.

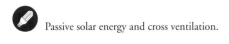
Passive solar energy and cross ventilation.

Natural materials: pine, mañío and ulmu wood.

The typical buildings of the local granaries and stables served as inspiration for the house design. The structure was built almost entirely in wood, except for specific metallic elements that were used to give the building greater stability. The house blends effortlessly into its leafy surroundings.

The interior of the house has an open plan layout. The wood cladding creates a warm, cozy atmosphere, in contrast with the humidity of the external surroundings. The metal and wood stairway suggests a lightness of structure, giving the feeling of more space in the communal areas.

The plans indicate the double-height spaces. In these common use areas, there is a greater exchange of air that cools the interior.

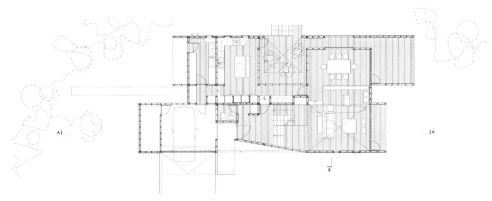

Ground floor

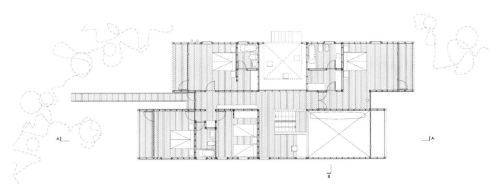

First floor

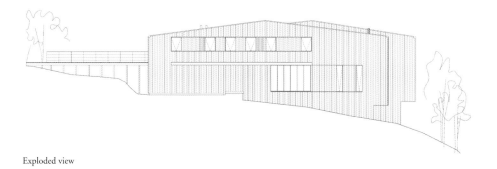

Exploded view

Longitudinal elevations

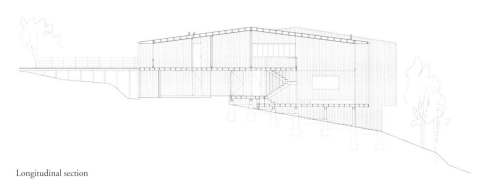

Longitudinal section

The elevations show how façades have more openings than others, this is because we have studied the orientation of the house to encourage natural light and ventilation. The section shows what is said in the text: the rear façade is very close to the forest, while the front façade enjoys views of the landscape.

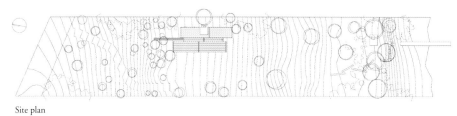

Site plan

The house is adapted to the terrain thanks to its different levels. The wooden bridge providing access to the property at the rear crosses over the uneven ground on this side of the building. Most of the windows are arranged around a semi-closed deck that has a good aspect to take full advantage of the sun in the early hours of the morning.

Restoration with environmental awareness

House in Bordeaux | Vincent Dugravier | Bordeaux, France | © Vincent Dugravier

This original restoration has converted an old neglected garage into a comfortable family home, taking advantage of some simple passive systems to create a sustainable building.

The original space consisted of two blocks: a warehouse with an area of 205 m² and a car port measuring 78 m². The architect decided to build a unique new volume that would take full advantage of the available sunlight. Savings are therefore achieved in lighting, and heat maximized. Furthermore, space was left for a garden that runs all the way round the house, oriented from north to south.

This façade, in translucent polycarbonate, allows for the direct recovery of the light that penetrates the house in the early hours of the morning and continues right round until sunset. When designing the house, cross ventilation was also taken into account, which is obtained thanks to the windows in the façade and the openings in the roof. Some of the materials used, such as the wooden frame and the outer walls, have been retained. The savings in materials and transport, both in terms of cost and CO_2 emissions, mean that reuse of part of the structure becomes an act of sustainability and environmental awareness. In keeping with this line of thought, other materials used have been sourced locally.

The rooms in the house are distributed as follows: on the top floor are the communal areas (living room, kitchen, dining room and study). These spaces are joined together in a single area, without any partitions, so the family spends most of their leisure time together. The bedrooms, bathrooms and dressing rooms have been arranged downstairs. Thus, all the rooms have direct access to the patio.

Despite the fact that the use of materials and the dimensions and height of the rooms respect a certain industrial style that impregnates the whole building, the decoration of the house is not at all cold. The choice of bright colors, modern furniture and the installation of parquet flooring upstairs has successfully created a warm atmosphere that is perfect for a young family.

 Use of sunlight, cross ventilation.

 Recycled materials that are locally sourced and adequate for maximizing hours of daylight.

The kitchen and dining room lead out to a veranda allowing the family to have their lunch outdoors when it is sunnier. The installation of translucent walls is one of the strong points in the house design, and was awarded the Architecture Prize for the City of Bordeaux.

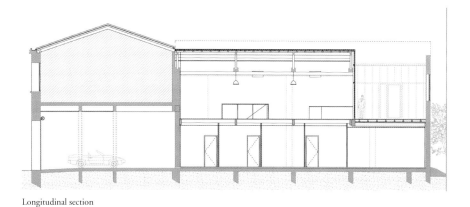

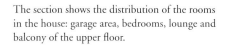
Longitudinal section

The section shows the distribution of the rooms
in the house: garage area, bedrooms, lounge and
balcony of the upper floor.

The old warehouse occupied the complete site.
Some new metal beams support part of the new
façade, which was created to take full advantage of
the natural light. The restored building is set within
the grounds of the former site occupied by the old
garage. A patio has been saved as a future garden
and play area for the children.

Restorations are also sustainable building systems. If some of the original building materials are used, savings are made in CO_2 emissions deriving from the manufacture and transportation of new materials. In addition, a lot of waste requiring treatment is not generated.

The floor plans show the outline of the old construction and the new areas. Part of the building has been converted into an outdoor patio.

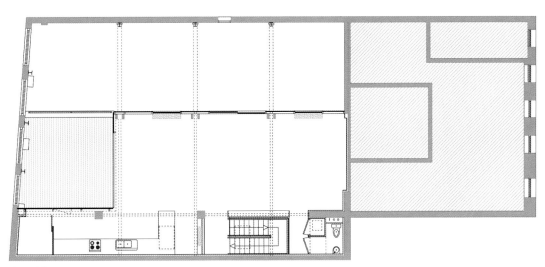

Ground floor

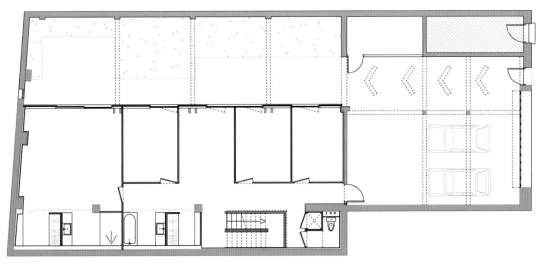

First floor

DIRECTORY

Beals Arquitectos
Chile
Tel. +56 2 881 8485
alejandrobeals@gmail.com

Belzberg Architects
1507 20th Street, Suite C
Santa Monica, CA 90404, USA
Tel. +1 310 453 9611
www.belzbergarchitects.com

Bligh Voller Nield
365 St. Pauls Terrace
Fortitude Valley
PO Box 801
Brisbane, QLD 4006, Australia
Tel. +61 7 3852 2525
www.bvl.com.au

Bricault Design
1395 Odlum Drive
Vancouver, BC V5L 3M1, Canada
Tel. +1 604 739 9730
info@bricault.ca
www.bricault.ca

Casey Brown Architecture
Level 1, 63 William Street
East Sydney, NSW 2010, Australia
Tel. +61 2 9360 7977
www.caseybrown.com.au

CCS Architecture
44 Mclea Court
San Francisco, CA 94103, USA
Tel. +1 415 864 2800
www.ccs-architecture.com

Coste Architectures
11, rue de la Prévôté - BP 19
78550 Houdan, France
Tel. +33 1 30 59 54 95
www.coste.fr

Dietrich Schwartz/GLASSX
Seefeldstrasse 224
8008 Zurich, Switzerland
Tel. +41 44 389 10 70
www.glassx.ch

Ecosistema Urbano Arquitectos
Estanislao Figueras, 6
28008 Madrid, Spain
Tel. +34 91 559 16 01
www.ecosistemaurbano.com

Francisco Portugal e Gomes
Rua de Vale Formoso, 602, 2º Esq.
4200-510 Porto, Portugal
Tel. +351 228 303 803
www.franciscoportugal.com

Franklin Azzi Architecture
2, rue d'Hauteville, 1er étage
75010 Paris, France
Tel. +33 1 40 26 68 21
agence@franklinazzi.com
www.franklinazzi.com

Hangar Design Group
Via Saffi, 26
20123 Milan, Italy
Tel. +39 024 802 8758
www.hangar.it

Marc Dixon
145 Russell Street
Melbourne, VIC 3000, Australia
Tel. +61 3 9663 6818
www.marcdixon.com

Olavi Koponen
Apollonkatu 23 B 39
00100 Helsinki, Finland
Tel. +358 9 441 096
www.kolumbus.fi/olavi.koponen

Patrick Marsilli
155 Voie Romaine
29000 Quimper, France
Tel. +33 2 98 57 60 60
www.domespace.com

Peter Kuczia
Osnabrück, Germany
Pszczyna, Poland
Tel. +49 163 929 50 50
www.kuczia.com

Ramírez Moletto
Nueva Costanera 4076
Vitacura, Santiago, Chile
Tel. +56 2 953 5248
www.ramirez-moletto.cl

Ray Kappe
715 Brooktree Road
Pacific Palisades, CA 90272, USA
Tel. +1 310 459 7791
www.kappedu.com
www.livinghomes.net

Rongen Architekten
Propsteingasse 2
41849 Wassenberg, Germany
Tel. +49 24 32 30 94
www.rongen-architekten.de

Stephen Taylor Architects
133 Curtain Road
London EC2A 3BX, United Kingdom
Tel. +44 20 77 29 16 72
www.stephentaylorarchitects.co.uk

Studio 804
Marvin Hall
1465 Jayhawk Blvd, Room 105
Lawrence, KS 66045-7614, USA
Tel. +1 785 864 4024
www.studio804.com

Tuomo Siitonen Architects
Veneentekijäntie 12
00210 Helsinki, Finland
Tel. +358 9 85 695 533
www.tsi.fi

UCArchitect
283 Lisgar Street
Toronto ON M6J 3H1, Canada
Tel. +413 536 4977
www.ucarchitect.ca

Una Arquitetos
Rua General Jardim 770 13ª
Vila Buarque, São Paulo, Brazil
Tel. +55 (11) 3231 3080
www.unaarquitetos.com.br

Vincent Dugravier/Dugravier + Sémondès
Architecture
43, rue Sullivan
33000 Bordeaux, France
Tel. +33 05 56 42 17 76
www.dugravier-semondes.com